D1503217

Ultimate Zero and One

Colin P. Williams
Scott H. Clearwater

Ultimate Zero and One

Computing
at the
Quantum Frontier

COPERNICUS
An Imprint of Springer-Verlag

© 2000 Colin P. Williams and Scott H. Clearwater

All rights reserved. No part of this publication may be reproduced, stored in a
retrieval system, or transmitted, in any form or by any means, electronic,
mechanical, photocopying, recording, or otherwise, without the prior written
permission of the publisher.

Published in the United States by Copernicus,
an imprint of Springer-Verlag New York, Inc.

Copernicus
Springer-Verlag New York, Inc.
175 Fifth Avenue
New York, NY 10010

Library of Congress Cataloging-in-Publication Data
Williams, Colin P.
Ultimate zero and one : computing at the quantum frontier / Colin P.
Williams, Scott H. Clearwater
p.   cm.
Includes bibliographical references and index.
ISBN 0-387-94769-8 (hardcover : alk. paper)
1. Quantum computers   I. Clearwater, Scott H.   II. Title.
QA76.889.W55   1998
004.1—dc21   98-42595

Manufactured in the United States of America.
Printed on acid-free paper.

9 8 7 6 5 4 3 2 1

ISBN 0-387-94769-8   SPIN 10537457

I dedicate this book to $(1/\sqrt{2})(|Zoe\rangle|Patricia\rangle + |Patricia\rangle|Zoe\rangle)$.

C.P.W.

To the Great Perfection that needs nothing extra to make it so.

S.H.C.

Contents

Preface

For 300 years physicists have been studying the forces of nature. It began with Newton's theory of gravitation and then continued in the nineteenth century with Faraday's and Maxwell's theories of electromagnetism and in this century with the strong and weak nuclear forces. It has even become fashionable to try combining all these forces into one Grand Unified Theory. Similarly, mathematicians have been formalizing the study of probability for 300 years and, in this century, the fields of algorithms, information theory, and computation. But only in the past decade has it become unmistakably clear to physicists and computer scientists that their respective fields are inextricably interwoven. Once a mere tool for physicists, computers must now be looked upon as ways and means of shedding light on fundamental questions of nature, such as whether nature itself is a computation. The ways and means are provided by the quantum computer, the first prototypes of which have been built in the past year.

Ultimate Zero and One is the most comprehensive book on quantum computation intended for the non-specialist. When we first began researching this book several years ago, there were a mere handful of theoretical research papers in existence and no experimental results. Now, new papers are appearing nearly every day, and several embodiments of quantum-mechanical switches have been built; the first specialized quantum computer was built and put into operation in early 1998. Even teleportation of a quantum state is now a reality. In just a few years, that scattered collection of papers has achieved a critical mass in what can now legitimately be called the field of quantum computing.

In *Ultimate Zero and One* we take the reader on a guided tour of the developments leading up to today's latest results. We begin by showing what we believe to be the inevitable march toward quantum computing, driven by the ever-decreasing size of computer components. Ultimately, the classical rules of logic and physics that have dominated the design of these components will no longer hold, and a new paradigm will prevail. That new paradigm is quantum computing—a new kind of programming logic and a new physical substrate for hardware. We will see that the neat logic of ones and zeros is replaced by a fuzzier notion of entangled oneness and zeroness. But it is precisely this cloudiness that leads to the fascinating capabilities of quantum computers—computers that not only can compute faster but also can compute what is uncomputable with today's computers.

In the first chapter, we explain why we believe quantum computing is the destiny of computers. Chapter 2 provides the reader with some background in the principles behind quantum computing. In Chapter 3 we explore how quantum computers can outperform today's classical computers. Chapter 4 discusses the first application of quantum computing—how quantum computers can be used to break codes with unprecedented speed. In Chapter 5 we talk about how the concept of randomness is intimately tied to the way the quantum world appears to operate and how we can harness true randomness only with a quantum computer. Capturing true randomness will make possible truer-to-life simulations of new pharmaceuticals and will find use in other applications as well. Whereas Chapter 4 reveals how quantum computers can break codes, Chapter 6 shows us how quantum computing can detect whether someone is eavesdropping on our secret messages so that we can change our code. Chapter 7 describes an application that sounds like the stuff of science fiction: teleportation. Not only is teleportation theoretically possible using quantum mechanics, but it has already been done! In Chapter 8 we consider the problem of maintaining the delicate conditions necessary to perform quantum computation reliably. In the final chapter we review a number of actual quantum computing devices and discuss the outlook for the future.

The first experimental apparatus for performing quantum key distribution—the basis for quantum cryptography. (Photograph by Robert Prochnow.)

Absent from *Ultimate Zero and One* is any discussion of the brain and consciousness. Certainly it is fascinating to speculate on such topics.[1] Indeed, there is no shortage of talk about whether brains are quantum computers and whether consciousness is a quantum phenomenon. But that is all it is—talk. Unfortunately, this plethora of talk is accompanied by a paucity of hard experimental evidence. There are hints of quantum phenomena going on in the brain, but no smoking gun has been discovered. Furthermore, the existence of quantum phenomena does not in itself imply that such phenomena are crucial for the functioning of the brain.

Still, it could be that quantum phenomena account for the late physicist David Bohm's observation:

Thus, thought processes and quantum systems are analogous in that they cannot be analyzed too much in terms of distinct elements, because the "intrinsic" nature of each element is not a property existing separately from and independently of other elements but is, instead, a property that arises partially from its relation with other elements.[2]

---

[1]Roger Penrose, *Shadows of the Mind* (Oxford, England: Oxford University Press, 1994).

[2]David Bohm, *Quantum Theory* (New York: Dover 1989), p. 169.

We hope that *Ultimate Zero and One* offers the reader a fascinating introduction to the computing machines of the future.

ACKNOWLEDGEMENTS

We would like to thank Allan Wylde for his foresight in signing us up for this project at a time when quantum computing was all but unknown. We also thank Copernicus editor Jerry Lyons for his infinite patience in pushing us to finish the book before quantum computers were actually built! Thanks also go to Steven Pisano for his attention to detail that made this project a quality production. Finally we thank Brian Oakley of the Bletchley Park Trust for providing information on code-breaking efforts during World War II.

One

⊰•◆•⊱

Computing at the Edge
of Nature

There's plenty of room at the bottom.
—Richard Feynman

The spread of computer technology into every aspect of modern civilization ranks as one of the greatest achievements in human history. The ability to access information at the touch of a button and to mix voice, image, and data in one torrent of information has revolutionized the way we communicate, solve problems, plan, shop, and even play.

Throughout this succession of technological advances, the foundations of the field of computer science have changed remarkably little. Contrast this with the revolutionary re-conceptualizations in biology and physics that have occurred over the same period. An understanding of the molecular basis for life has spawned a new biotechnology. Advances in quantum physics have led to high-temperature superconductors, atomic force microscopes for manipulating individual atoms, and plans for astronomical interferometers to take high-resolution pictures of planets orbiting distant stars. Can computer science escape the implications of the advances in these other fields? We believe it cannot.

Over the past 40 years there has been a dramatic miniaturization in computer technology. If current trends continue, by the year 2020 the basic memory components of a computer will be the size of individual atoms. At such scales, the mathematical theory underpinning modern computer science will become invalid. A new field called *quantum computing* is emerging, which is forcing scientists to reinvent the foundations of computer science and information theory in a way that is consistent with quantum physics—the most accurate model of reality that is currently known.

The product of this endeavor will be the concept of a "quantum computer"—a physical device that utilizes subtle quantum effects in an essential way, and whose natural evolution over time can be interpreted as performing a desired and controllable calculation.

Remarkably, the new computer science predicts that quantum computers will be able to perform certain computational tasks in phenomenally fewer steps than *any* conventional ("classical") computer. Moreover, quantum effects allow unprecedented tasks to be performed, such as teleporting information, breaking supposedly "unbreakable" codes, generating true random numbers, and communicating with messages that expose the presence of eavesdropping. These capabilities are of significant practical importance to banks and government agencies. Indeed, a quantum scheme for sending and receiving ultra-secure messages has already been implemented over a distance of 30 kilometers—far enough to wire the financial district of any major city (Marand, 1995).

Such a revision in the foundations of computer science does not imply, however, that classical computers will become obsolete. By analogy, although Einstein's theory of relativity provides a deeper understanding of physical reality than that supplied by Newtonian physics, Newtonian physics is still *good enough* to predict the behavior of many practical physical systems. Similarly, classical computers will remain good enough for solving many practical computational problems. However, there will be certain specialized computational problems that can be solved more efficiently using quantum computers rather than classical ones. These are the kinds of problems that will drive the development of quantum computers.

Although modern computers already exploit some quantum phenomena, such as quantum tunneling in transistors, they do not make use of the full repertoire of quantum phenomena that nature provides. Harnessing the more elusive quantum phenomena will take computing technology to the very brink of what is possible in this universe—indeed, to the very edge of nature!

Rethinking Computers

When you think about computers, you probably picture the helpful device humming gently on the top of your desk. However, in reality, "computers" are much more pervasive in the world around you. In fact, some people believe that we can think of the entire universe as a computer whose "program" ultimately dictates the observed laws of physics.

In the nineteenth century the word *computer* meant a person who had been trained to perform routine calculations using little more than quill pen, ink, and paper. Such people did not need much education because their tasks were simple and repetitive. They were selected instead for personality traits such as perseverance and meticulousness.

Today, our idea of a "computer" is rather different. The term conjures up an image of a box of high-tech electronics whirring with frenzied activity; gobbling up diskettes chock full of the latest software; firing up word processors, spreadsheets, or games; and ejecting page after page of pristine laser-printed output. At least this is the image of computers in our offices and homes.

Supercomputers, the heavyweights of the computer world, are serious machines for predicting the weather, searching vast databases, updating company records, and solving scientific problems. But in all important respects, supercomputers are entirely equivalent to the more humble computers we deal with every day. Granted, they may be much faster and may be equipped with a gargantuan memory, but they cannot solve a single problem that our personal computers cannot also solve, given enough time and memory.

However, the image of computers is changing; they are starting to blend into the woodwork by becoming "embedded." Examples are already around us. The laser scanner at the supermarket checkout is hooked up to a computer that tracks inventory. The ATM at the bank dispenses cash and debits your account. The latest cars contain computers that determine when to shift gears and when to fire the spark plugs. Computers are starting to disappear before our very eyes by becoming part of the fabric of our world.

These multiple meanings of *computer,* from human being to electronic assistant to embedded artifact, actually point to there being something much deeper to the concept. In all these cases the "computer" is really a physical device that takes some input, transforms it, and produces some output. For example, at the supermarket checkout, the laser scans the black-and-white bar code on a product to produce a sequence of 1s and 0s (called *bits,* for **bi**nary dig**its**) that are then sent to the computer. The computer transforms the sequence of numbers into the name of the product being purchased, then converts the name into the price, and finally prints it out on a receipt.

At a minimum, a computer is a physical device that makes possible such transformations between patterns, or "computations," as they are called. When the patterns take on a continuum of values it is an analog computer. When they take on a discrete set of values it is a digital computer. Moreover, regardless of the actual symbols used (such as the dollar prices of various shopping items), inside a computer all the symbols are encoded into long sequences of 0s and 1s (such as 1011010101011111-01011011). The computation is then performed by transforming, over what may be many steps, the input sequence of 0s and 1s into some other sequence of 0s and 1s. Finally, the output is created by decoding the final sequence of 0s and 1s back into meaningful symbols (such as "Total = $52.47"). The computations don't have to be purely arithmetic, however. By generalizing the concept of computation to encompass arbitrary transformations of symbols, we can recognize all sorts of tasks as being "computations" and hence amenable to automation. For example, the ancient Greeks invented a rule of inference that could be expressed symbolically as the following triplet of statements: "X is true" and "if X is true then Y is true," from which one could conclude that "Y is true." An elaboration of these ideas is what enables computers programmed using artificial intelligence techniques to mimic logical reasoning. Thus, procedures that were once regarded as requiring intelligence can be couched as purely symbolic computations.

Surprisingly, this is not at all a recent idea. As early as the seventeenth century, Baron Gottfried von Leibniz expressed the hope that someday all philosophical disputes would be resolved by the parties huddling around a mechanical reasoning contraption, configured to reflect the logical positions of all the participants in the dispute and set in motion with the resounding decree "Gentlemen. Let us compute!"

The essence of a computer, then, boils down to just three basic ideas: (1) A computer is first and foremost a physical device. (2) It can be thought of as containing rules for transforming some initial pattern of 0s and 1s into some final pattern of 0s and 1s. (3) When suitably interpreted, the sequence of steps in this transformation process can be understood as the execution of some desired computation. Current computers use classical electromagnetic effects to accomplish the necessary transformations. However, as we shall see, this need not be the case. With smaller components, other physical effects might become practicable.

Shrinking Technology

So far the successive improvements in the miniaturization of computer technology have been due to ever more ingenious employment of the physical opportunities nature has provided. For example, the machines of the nineteenth century used gear trains to perform their computations; in the 1940s and early 1950s, computers used vacuum tubes; in the 1950s and 1960s, transistors appeared; these were quickly followed by integrated circuits in the 1970s and finally by microchips. Each stage in this progression was marked by the discovery of how to map the basic operations of a computer onto smaller and smaller physical devices. Thus, whereas gears use roughly a trillion trillion atoms to represent one bit, microchips have succeeded in economizing down to merely a billion or so. Such exploitation of nature's resources is impressive but still extravagant in comparison to the ultimate limit of computers. By extending the trend of miniaturization, it is not difficult to envision machines that will use a single atom to represent a single bit—that is, a 0 or a 1.

Because computers are physical devices, it is natural to think about the space, time, and energy implications of trying to make computers faster. To make computers faster, their components have to be squeezed closer together, because the signals that need to be passed around inside the computer cannot travel faster than the speed of light—roughly 186,000 miles per second in a vacuum. The closer components are together, the less the time it takes for them to communicate. One way to pack more components into a given volume is to make the components smaller. This entails finding smaller physical systems that can function as well as those they replace. This is what happened when vacuum tubes were replaced by transistors and transistors by integrated circuits.

However, size is not the only issue. We also need to make the components of the computer work faster—that is, perform more logical operations per second. This improves the efficiency of the chip. Unfortunately, the components inside conventional computers give off a certain amount of heat as a by-product of their operation. If we simply pack the components closer together and run them at a higher rate without also improving their energy efficiency, the computer will melt as it computes! Already supercomputers need a liquid refrigerant cycled through their circuitry to cool them down. Consequently, not only do the basic operations have to be mapped onto smaller physical processes but they also have to be made more energy-efficient. Solving either of these problems helps; solving both will be essential to achieving further miniaturization and speedup.

Consider a representative measure of computer miniaturization: the trend in the number of atoms needed to encode one bit of information. In the 1970s, Gordon Moore, a co-founder of Intel, noticed that the memory capacity of a computer chip doubles approximately every year and a half even though chips remain, physically, about the same size (Malone, 1995). This means that every 18 months, only half as many atoms are needed to store a bit of information. Robert Keyes of IBM has analyzed the number of atoms needed to represent one bit of information as a function of time and has represented this information graphically. In Figure 1.1, the vertical axis uses a special compressed "logarithmic" scale, which is ideal for plotting a quantity that varies vastly in magnitude. A number of the form 10^n means 10 multiplied by itself n times. If you look closely, you will see that the number of atoms needed to encode a bit of information decreases by about a factor of 10 every 5 years.

From this figure we can see why, in 1959, Nobel laureate Richard Feynman was inspired to entitle his address to the American Physical Society as "There's Plenty of Room at the Bottom" (Feynman, 1960). Despite the breathtaking size reductions of the last three decades, today's computer chips are still about 100 million times bigger than the one-atom-per-bit level. Feynman was extraor-

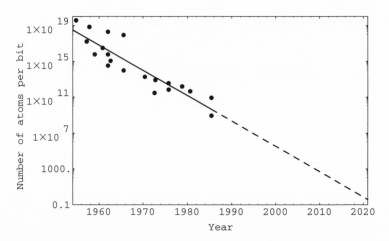

Figure 1.1   The number of atoms needed to represent one bit of information as a function of calendar year. Extrapolation of the trend suggests that the one-atom-per-bit level will be reached in about the year 2020. Adapted from Keyes, 1988.

dinarily foresighted; there is *still* plenty of room at the bottom—
that is, at the tiniest scales of matter. We can see that the trend,
which has held since 1950, reaches a limit of one-atom-per-bit
around the year 2020. At the one-atom-per-bit level, it will be *necessary* to use quantum effects to read bits from, and write bits to, the
memory register of an ultra-small computer. Thus, even on memory
grounds alone, there is a strong reason to investigate the operating
principles and feasibility of quantum computing devices.

Similar trends can be plotted for the increase in the number of
transistors per chip and the increase in the speed at which chips can
be operated. An extrapolation of the trend in clock speed suggests
that the computers available in 2020 will operate at about 40 GHz
(40,000 MHz). By comparison, Intel's Pentium-III processor, re-
leased in 1999, operates at a speed of 500 MHz. Similarly, if we ex-
trapolate the trend in energy needed to perform one logical operation would decrease to that of typical atomic energy scales by about
2020. Thus, if current trends in miniaturization and energy efficiency continue as they have for the past 30 years, computer technology should arrive at the quantum level by about the year 2020.
At this point, every aspect of computer operation, from loading data
and running programs to reading memory registers, will be dominated by quantum effects.

Remarkably, far from fearing that such ultimate miniaturization
will mark the end of the road for computer development, scientists
are beginning to realize that it may be possible to *harness* quantum
effects rather than merely contend with them. This is a subtle but
important shift in perspective. The resulting "quantum computers"
would exploit quantum effects to perform feats of computation and
communication that are utterly impossible with any computer that
obeys the laws of classical physics, no matter how advanced it may
be. Even a massively parallel, bit-guzzling behemoth of a supercomputer is no match for a quantum computer on certain computational tasks.

However, we are not likely to *begin* the final computer revolution by producing a top-of-the-line quantum computer. The first
quantum computers will be extremely rudimentary. It would be un-
fair to expect more: The current cumulative investment in quantum
computing research is far less than $50 million. By comparison, our
current (classical) computers represent a $1 trillion investment.
Comparing a state-of-the-art classical supercomputer with a prototype quantum computer is like comparing a Ferrari sports car to a
hot air balloon. The Ferrari marks the pinnacle of engineering excel-

lence along one research dimension. The result is a machine that is superbly suited for operation on a two-dimensional surface or, more accurately, a one-dimensional ribbon of tarmac. By contrast, the hot air balloon is a first attempt to strike out in a completely new direction—in this case, a new dimension ("up"). Hot air balloons seem slow and cumbersome compared with Ferraris, but they are vastly more versatile.

In the next section we begin to outline the quantum analog of the "up" dimension for our hot air balloon. That is, we describe some of the phenomena that are available to a quantum computer but not to a classical computer. It is ultimately the ability to exploit novel physical phenomena (new "dimensions") that gives quantum computers the edge over classical computers.

A Peek Into Quantumland

What is so special about operating a computer at the quantum scale? To understand this, we need to take a look at the strange world of quantum physics, a world teaming with ghostly virtual particles, entangled interactions, and deep metaphysical conundrums.

By the end of the nineteenth century, physicists were fairly content with their understanding of nature. It seemed that the next generation of physicists was destined to tidy up the theories of their predecessors, repeat a few experiments to a higher precision, derive results more elegantly, and start looking for another, more fruitful line of work!

However, the triumph of classical physics was short-lived. Its demise could have been foreseen in the great foundries that fueled the Industrial Revolution. As all ironworkers know, when you take a lump of iron and heat it up, it starts to glow red. Heat it some more and it glows orange-yellow. Understanding why iron glows red turns out to require new physics—*quantum* physics.

The view of classical physicists was that light was a kind of electromagnetic wave motion. An electromagnetic wave consists of oscillating electric and magnetic fields that are at right angles to one another and to the direction of motion. They knew that the color of light was related to the frequency of these electromagnetic waves. The fact that the color of an object changed when it was heated led them to speculate that the things emitting the light vibrated faster and faster as they were heated. But according to classical physics, the tiny cavities inside the metal have room for many more short-wavelength light waves than long-wavelength light waves, so energy should accumulate in the short wavelengths. Because wavelength and frequency are inversely related, a hot object, such as a lump of molten iron, ought to glow with short-wavelength "blue" light (or actually *beyond* blue, in the ultraviolet) rather than red light. Nobody could explain how to get classical physics to predict a red glow instead.

Max Planck, a German physicist, had been working on the problem of hot, glowing objects. He discovered that he could explain the observed spectrum of a hot body if he assumed that the energy given off from an object was proportional to a multiple of the frequency of vibration of the objects emitting the light: $E = hn\nu$, where E is the energy, n is the multiple, ν is the frequency, and h is a constant. Planck found that when he set h to zero, his theory gave the same faulty prediction as classical physics—that hot objects glow ultraviolet. But when he set h to a particular, but very small, value equal to $h = 6.6256 \times 10^{-34}$ Joule seconds, his theory gave the correct prediction.

What Planck had stumbled onto was the first great idea of quantum mechanics—that certain physical quantities, such as energy, cannot take on a continuum of possible values but must instead appear in multiples of Planck's mysterious constant h, the so-called "quantum of action."

We now know that Planck's original proof for why hot objects glow red contained a flaw. Nevertheless, his idea of quantizing certain physical quantities was the critical step in developing the new quantum mechanics. Once the idea of quantization was known, other puzzling experimental results could be explained readily. These included Einstein's explanation of the photoelectric effect and Bohr's explanation of the lines in atomic spectra. The photoelectric effect was particularly important because it led to the second great idea of quantum mechanics—that particles can behave as waves and waves can behave as particles, depending on the experimental arrangement.

The photoelectric effect describes the emission of electrons from the surface of a metal that is illuminated with light. Experimentally, the energy of the liberated electrons is found to depend on the frequency of the illuminating light but not on its intensity, or brightness. This phenomenon is puzzling if, as classical physics asserts, light is a kind of wave.

According to classical physics, the energy imparted by a wave is proportional to its intensity, that is, the square of its crest-to-trough distance. So, if light is a kind of wave, then higher-intensity light ought to liberate higher-energy electrons. But this is not what is observed experimentally. Instead, higher-intensity light causes more electrons to be liberated, but they each are found to have the same energy as before. The energy is set by the wave's frequency (i.e., color) rather than its intensity.

Albert Einstein's explanation for the photoelectric effect was that light is not, after all, a wave but instead must consist of tiny particles, called photons, each of which carries a lump of energy given by the Planck formula $E = h\nu$. An electron is liberated from the surface of a metal whenever a photon strikes it and transfers one whole quantum's worth of energy. If the metal is illuminated with a brighter light, more electrons are liberated per second, but they will all have exactly the same energy. Thus the photoelectric effect is an experiment in which supposed "waves of light" behave more like "particles of light."

Conversely, other experiments showed the exact opposite: Under the right circumstances, objects conventionally thought of as particles can behave instead as though they are waves. For example, if you repeatedly dab a ruler into a tank of water, you will set up a series of parallel water waves that move away from the ruler. If you position in the path of these waves a barrier that contains two vertical slits, the waves will pass through the slits and appear on the other side as two sets of concentric waves, each centered on one of the slits, that soon run into one another. If you look closely, you will see that wherever two wave crests coincide, you get a bigger crest and that wherever two wave troughs coincide, you get a bigger trough. But when a crest of one wave coincides with a trough of the other, you get no net disturbance of the water surface. This pattern of disturbance, in which waves add up at certain places and cancel out at others is called *interference,* and it is the hallmark of wave-like behavior.

It happens that we can perform a similar experiment with electrons. In classical physics, electrons are thought of as tiny particles. Particles have a well-defined location, typically move in straight lines, and bounce off other particles much as billiard balls do. Consequently, if you fired an electron at a pair of slits in a metal foil and looked to see where the electron landed on the far side of the foil, you might expect to find the electron in one of two locations, given by projecting a straight line from the electron source through each slit to the recording screen. However, if you do the experiment, this is not what you see. Just as in the case of water waves, the pattern you see with the electron detector has several bright and dark fringes that correspond to where the electron is found and to where

it is not found, respectively. The pattern can be explained trivially if
you describe the electron as a wave having a wavelength equal to
$\lambda = h/p$, where p is the electron's momentum and h is (you guessed
it) Planck's quantum of action.

Thus, everyday experiences do not foster reliable intuitions
about the true, quantum nature of reality. Events that seem impossible at our scale can and do happen at the quantum scale. They are
no less real simply because we humans find them perplexing. Careful experiments on minute physical systems, such as photons of light
and electrons, have revealed that, depending on the kind of experiment used, such systems can at one moment appear more particle-
like and at another more wave-like. Sir William Henry Bragg, an
early experimenter with quantum phenomena, once described the
Jekyll-and-Hyde personality of physical systems by joking, "Physicists use the wave theory on Mondays, Wednesdays, and Fridays,
and the particle theory on Tuesdays, Thursdays, and Saturdays."
But it's really no joke. At sufficiently small scales, things we are used
to thinking of as particles can behave as waves, and things we are
used to thinking of as waves can behave as particles. This duality is
very important in quantum computation. Different computations
going on simultaneously inside a quantum computer can be made to
interfere with one another to yield a net result that reveals some
joint characteristic of all the simultaneous computations.

Since these early successes, the theory of quantum physics has
been subjected to intense scrutiny by the scientific establishment.
Remarkably, experiment after experiment has confirmed the predictions of quantum physics, no matter how preposterous they might
have seemed. Yet quantum physics remains an enigma. It asks us to
abandon many of our most cherished assumptions about the nature
of reality. It is often at odds with what most people regard as common sense, but it seems to provide exactly the right language for describing the operations of ultra-small computers. So what is this
new quantum physics all about? What kinds of things does it deal
with and talk about? How does it describe physical reality?

Quantum physics attempts to answer three basic questions:
"How do you describe the state of a physical system?" "How does
the state change if the system is not observed?" and "How do you
describe observations and their effects?" The strangeness of quantum mechanics lies not so much in the questions it asks but rather in
the answers to those questions that it provides.

First, the description of the state of a physical system is given in
terms of what is called a *state vector*. Mathematically, this is a col-
umn of numbers that can include "complex" numbers that have

both a real component and an imaginary (i.e., a multiple of $\sqrt{-1}$) component. These numbers, called, *amplitudes,* are related to the probability of finding the physical system in a particular state, or configuration, if we were to observe it.

Although it is hard to visualize, we can picture the state vector as an arrow in a space spanned by a set of mutually perpendicular axes. The axes can correspond to the configurations in which it is possible to observe the system in the sense that, whenever we do observe the system, we project the state vector arrow to lie along just one of the axes that span the space.

The state vector contains *complete* information about the quantum system in the sense that, given the state vector, we can calculate the expected value of *any* observable property of the system. This is really quite remarkable; somehow the state vector must encapsulate everything of significance about the quantum system.

Quantum mechanics says that there are two qualitatively distinct kinds of dynamical processes that describe how a quantum system changes over time. If the system is not being observed, then its state changes smoothly and continuously in accordance with a special differential equation discovered by Austrian physicist Erwin Schrödinger in 1926.

However, if the quantum system *is* observed, then the smooth, continuous evolution appears to be interrupted abruptly, the system seems to jump into one of a select number of special states called *eigenstates,* and the result of the measurement is a corresponding real number called an *eigenvalue.* Thus, eigenstates can be thought of as "states you are allowed to find the system in," and eigenvalues as "the numbers that come out of measuring instruments indicating which eigenstate the system is in." However, when we make a measurement, there is no way to predict, with certainty, *which* eigenstate will be selected. The best we can do is to characterize the relative probability of each possible outcome.

As you can see, the answers that quantum mechanics provides to the questions of state description, state evolution, and measurement are fundamentally different from what you might expect on the basis of everyday experiences. In the everyday world, most quantities that we can measure appear to vary over a continuous range of values. But in the quantum world, the only allowed values are the so-called eigenvalues, which typically constitute a discrete set. Moreover, when we look at something in the everyday world, we do not expect to change or perturb it in any way whatsoever. But in the quantum world we invariably perturb the state of the system whenever we look at it. What is worse, we cannot even predict with certainty what answer we will get. Quantum physics forces us to make do with probabilistic predictions.

Thus, you should not feel uneasy if you find quantum physics hard to grasp: It requires a major revision in your assumptions about the nature of physical reality. In fact, Neils Bohr, one of the founders of quantum theory, once said, "Anyone who thinks he understands quantum theory doesn't really understand it." Much of the difficulty with quantum theory lies in how we are to *interpret* what it is telling us about the fundamental nature of reality. An "interpretation" is a story that we tell about what is *really* going on behind the equations. Although the nature of reality inspired many great minds of the twentieth century to propose competing accounts, "reality" research has fallen out of favor with mainstream physicists today. This is partly because it is very difficult to conceive of experiments that can refute any of the surviving interpretations and partly because all of the interpretations make the same quantitative predictions. More important, perhaps, the mathematical machinery of quantum physics works so well in practice that there is little perceived need for philosophical inquiry. In fact, if most physicists adhere to any interpretation of quantum theory, it might best be called the "Let's shut up and calculate" interpretation!

Nevertheless, we have to hang our quantum hat on some interpretation, so we will implicitly assume what is called the Copenhagen Interpretation. This very old interpretation, developed by Neils Bohr and his colleagues in Copenhagen, is still surprisingly popular today. Indeed, the physicist Anton Zeilinger, who developed the first experimental demonstration of quantum teleportation, a very *modern* idea, still sees himself as a Copenhagenist. We are in good company.

The question that concerns us next is how to relate the phenomena of quantum physics to the elementary operations of a computer.

The Qubit: Ultimate Zero and One

In any digital computer, each bit must be stored in the state of some physical system. Contemporary computers use voltage levels to encode bits. Old-fashioned, mechanical computers use the position of gear teeth. The only requirement is that the physical system must possess at least two clearly distinguishable configurations, or states, that are sufficiently stable that they do not flip, spontaneously, from the state representing the bit 0 into the state representing the bit 1, or vice versa.

Fortunately, certain quantum systems possess properties that lend themselves to encoding bits as physical states. When we measure the "spin" of an electron, for example, we always find it to have one of two possible values. One value, called "spin up" or $|\uparrow\rangle$, means that the spin was found to be parallel to the axis along which the measurement was taken. The other possibility, "spin down" or $|\downarrow\rangle$, means that the spin was found to be anti-parallel to the axis along which the measurement was taken. This intrinsic "discreteness," a manifestation of quantization, allows the spin of an electron to be considered a natural "bit." As a matter of fact, there is nothing special about spin systems. Any two-state quantum system, such as the direction of polarization of a photon or the discrete energy levels in an excited atom, would work equally well. Whatever the exact physical embodiment chosen, if a quantum system is used to represent a bit, then we call the resulting system a quantum bit, or just "qubit" for short.

Inasmuch as we are talking about both bits and qubits, we'd better find a way of distinguishing them. When we are talking about a qubit (a quantum bit) in a physical state that represents the bit value 0, we'll write the qubit state as $|0\rangle$. Likewise, a qubit in a physical state representing the bit value 1 will be written $|1\rangle$.

For reasons that will become apparent shortly, it is useful to picture the state of a qubit as a point on the surface of a sphere (see Figure 1.2). We can impose the arbitrary convention that the north pole, or "spin-up" state, represents the bit 0 and that the south pole, or "spin-down" state, represents the bit 1.

However, because a qubit is a quantum system, it is governed by the laws of quantum physics, not classical physics. In quantum

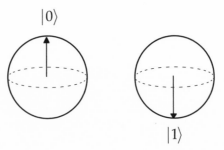

Figure 1.2   A qubit representing the bit 0 (left) and the bit 1 (right).

physics, if a quantum system can be found to be in one of a discrete set of states, it may also exist in a *superposition*, or a blend of some or all of those states *simultaneously*. Thus, whereas at any instant a classical bit can be *either* a 0 *or* a 1, a qubit can be a superposition of *both* a $|0\rangle$ *and* a $|1\rangle$—that is, a quantum state of the form $c_0|0\rangle + c_1|1\rangle$, for which the sum of the squares add up to one: $|c_0|^2 + |c_1|^2 = 1$. Here the numbers c_0 and c_1 are typically "complex" numbers. A complex number has both a real part, x, and an imaginary part, y, such that $c_0 = x_0 + iy_0$ and $c_1 = x_1 + iy_1$, where $i = \sqrt{-1}$ and x and y are both real ("ordinary") numbers. The squares of the weighting factors of 0-ness and 1-ness are related to the real part and the imaginary part of the complex numbers via the equations $|c_0| = \sqrt{x_0^2 + y_0^2}$ and $|c_1| = \sqrt{x_1^2 + y_1^2}$. Note that the proportions of 0-ness to 1-ness need not be equal. Any combination is allowed, provided that the sum of the squares is one. Pictorially, the arrow representing the state of a qubit is a superposition state may point to anywhere on the surface of the sphere (see Figure 1.3).

Unfortunately, if we do not know what state—that is, what superposition state—a qubit is in, when we attempt to "read" or "measure" or "observe" it, we are likely to disturb its state irrevocably. To make the measurement, we need to know what kind of physical system embodies the qubit. If the qubit is embodied as the spin of an electron, then we can measure the direction of this spin relative to some axis by passing the electron through a magnetic field aligned with this axis and seeing which way the electron is deflected. The "spin" of an electron interacts with a magnetic field in a predictable way. If the qubit encodes a $|0\rangle$, for example, the electron is deflected upward by the magnet. If the qubit encodes a $|1\rangle$, it is deflected downward by the magnet. The problem comes when the qubit encodes a superposition state such as $c_0|0\rangle + c_1|1\rangle$. In this

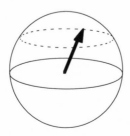

Figure 1.3   A qubit is a blend of both a 0 and a 1, but the contributions of 0-ness and 1-ness need not be equal.

case, we cannot predict with certainty whether the qubit will be deflected upward or downward. However, we can give the probability of occurrence for each of these events. A *probability* is a number between 0 and 1 that is a measure of how likely an event is to happen. A probability of 0 means that the event certainly will not happen, and a probability of 1 means that the event certainly will happen. If the qubit is in the state $c_0|0\rangle + c_1|1\rangle$ and is passed through our qubit-reading apparatus, it will be observed to be in the $|0\rangle$ state with probability $|c_0|^2$ and it will be observed to be in the $|1\rangle$ state with probability $|c_1|^2$. No other outcome is possible. Moreover, at the instant the measurement is made, the qubit is projected into the state that is compatible with the observed outcome. Thus if you find the qubit in state $|0\rangle$, then immediately after you have made this observation the qubit will indeed *be* in state $|0\rangle$, even if it was not in this state before the measurement was made. This illustrates that measurements made on quantum systems that are in superposition states invariably perturb the system being measured in a way that cannot be predicted. This is an important point. If no record is kept of the result of a measurement; then the measurement can be undone. But if a record is kept, the measurement is irreversible.

Before we leave the qubit, we should address one final issue that will prove to be of vital importance in quantum computation. We want to explain why we drew, or rather *had to draw,* the state of the qubit as an arrow in a *sphere* rather than an arrow in a *circle.* The reason is that there are many ways of choosing a pair of (complex) numbers c_0 and c_1 so that the qubit has the same relative proportions of 0-ness to 1-ness. The parameter that distinguishes between these possibilities is the angle about the vertical axis piercing the north and south poles of the sphere, through which the arrow representing the qubit state is rotated. This parameter is called the *phase,* and it has no analog for classical bits (see Figure 1.4).

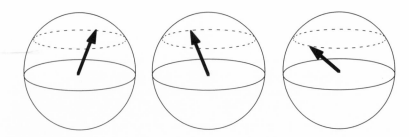

Figure 1.4 Three qubits that all have the same proportions of 0-ness to 1-ness but have different phases.

Phase becomes important when we add together the effects of a computational operation on all of the terms in a superposition. For example, consider a pair of qubits; let's call them A and B. Qubit A has been prepared in the state $|\psi\rangle = c_0|0\rangle + c_1|1\rangle$ and qubit B has been prepared in the state $|\phi\rangle = c_0|0\rangle - c_1|1\rangle$. These two quantum states are similar in terms of amplitudes, as the relative proportions of 0 and 1 are the same in both cases. Thus, if you were to measure the bit-value of qubit A, you would find it to be 0 with probability $|c_0|^2$ or 1 with probability $|c_1|^2$. Likewise, if you were to perform a similar measurement on qubit B, you would also find it to be 0 with probability $|c_0|^2$ or 1 with probability $|c_1|^2$ because $|-c_1| = |c_1|$. However, the states $|\psi\rangle$ and $|\phi\rangle$ differ in the *sign* with which the $|1\rangle$-component appears. Hence, they are said to differ by a *phase* factor. This might seem insignificant given that the probabilities of the two possible measurement outcomes are the same. However, the phase factor becomes important when we use the qubit as an input into a quantum computer and run some quantum algorithm upon it.

For example, suppose you feed either qubit A or qubit B into the same quantum computer, perform some quantum algorithm, and read the bit-value of the output. Does it make a difference whether we use qubit A or qubit B given that the relative proportions of the $|0\rangle$-component and $|1\rangle$-components are the same? To be concrete, let's choose a particular quantum computation to be performed. Let the computation be defined according to the following rule: if the input state has the form $a|0\rangle + b|1\rangle$, then the output state has the form $\left[(a+b)/\sqrt{2}\right]|0\rangle + \left[(a-b)/\sqrt{2}\right]|1\rangle$. Thus, given that the input qubit is in the state $a|0\rangle + b|1\rangle$, upon performing the indicated computation and reading the bit-value of the output we will obtain 0 with probability $|(a+b)/\sqrt{2}|^2$ or 1 with probability $|(a-b)/\sqrt{2}|^2$.

Now compare what happens when we run this quantum computation on first qubit A and then qubit B as input. Using qubit A, the input state is $|\psi\rangle = c_0|0\rangle + c_1|1\rangle$ and the output state is $\left[(c_0+c_1)/\sqrt{2}\right]|0\rangle + \left[(c_0-c_1)/\sqrt{2}\right]|1\rangle$. Using qubit B, the input state is $|\phi\rangle = c_0|0\rangle - c_1|1\rangle$ and the output state is $\left[(c_0-c_1)/\sqrt{2}\right]|0\rangle + \left[(c_0+c_1)/\sqrt{2}\right]|1\rangle$. A special case of the inputs occurs when $c_0 = c_1 = 1/\sqrt{2}$. In this case, the final answer that our quantum computer spits out, using qubit A as input, will surely be $|0\rangle$. Conversely, the final answer, using qubit B as input, will surely be $|1\rangle$. But qubit A and qubit B differ only in the *phase* with which the $|1\rangle$-component appears. Hence, this example shows that the final result of a quantum computation can be acutely sensitive to the phase factors in the input state.

Are Bits Driving Us Bankrupt?

It is becoming more and more costly for chip manufacturers to maintain the pace of Moore's Law (see the "Shrinking Technology" section in this chapter). Despite competitive business pressures driving efficiency, the cost of a new semiconductor plant has been doubling every 3 years (Hutcheson, 1996). In 1994, Motorola proposed building a $2.4-billion plant. If the triannual doubling continues, by 2020 the cost of building a semiconductor plant will be $1 trillion, which will be over 5 percent of the projected gross domestic product of the entire United States! Clearly, this economic trend is not sustainable. To be profitable, Motorola would need sales in 2020 to be $10 trillion (over half the predicted GDP). If computer performance is to be sustained, it looks like we will have to move to some new chip technology that can deliver the desired improvements in performance and can be mass-produced more economically. Could this technology be quantum computing?

As we shall see later in the book, quantum computers are predicted to be significantly more efficient at certain computational tasks than conventional computers. Moreover, quantum phenomena allow entirely new kinds of tasks to be performed, including faithful simulations of molecular processes, super-dense information coding, and ultra-secure communication. It is too soon to tell whether our computing needs in 2020 will be dominated by the kinds of tasks at which quantum computers excel or, indeed, whether they can be manufactured cheaply. Nevertheless, they offer a distinct and relatively unexplored alternative to conventional computer technology that may have unforeseen financial payoffs. Certainly it is conceivable that there might be a market for highly specialized quantum computers that perform vital, gargantuan computations such as drug design and cell biology simulations.

Apart from the raw economics of building computers, there is also the cost of running them. Currently, 5 percent of the total power generated in the United States is consumed by computers (Malone, 1995). Assuming that the total number of computers remains constant, and extrapolating from current trends in clock speed, memory capacity, and energy efficiency, we estimate that the computers of 2020 will operate at 40 GHz, have 160 GB RAM, and dissipate about 40 W of power. This is comparable to the power requirements of current machines and would therefore still consume roughly 5% of the current total power generation capacity. Unfortunately, it is hard to believe that the number of computers will not grow. At 40 W of power consumption, with the cost of building a new power plant so high, fossil fuels continuing to dwindle, fission power in disfavor with the public, and fusion power still many decades away, the drain that computers inflict on our power supply could become significant.

In this regard, quantum computers might again offer an advantage. Theoretically, any energy you need to feed into a quantum computer to do a particular computation should be redeemable at the end of the computation, resulting in no net energy consumption. The only time you need to expend energy is to erase the contents of a memory register. In practice, it is most unlikely that such a perfect energy balance will be achievable, at least if you want the computer to work at a reasonable speed. A more realistic advantage of a quantum computer over any classical computer (no matter how advanced) is that certain types of computation can be done using far fewer logical operations on a quantum computer than on a classical computer. If you can reduce the number of logical operations to achieve a given result, you might be able to save on the energy required for the computation.

Thus, there appear to be several advantages of moving computation to the quantum level. The economic trends suggest that current computer technologies will not remain cost-effective indefinitely. A technological paradigm shift will be necessary at some point if we are to continue to meet the growing demand for raw computational power. This may or may not be quantum computing, of course. It is possible that superconducting electronics circuits or some other technology will offer a more cost-effective approach to classical computing (Likharev, 1996), but even so, advanced classical computers such as those based on superconducting electronics, may only offer a temporary stopgap.

The engineering trends suggest that miniaturization of computer components will attain atomic proportions by around 2020. If this trend holds true, we really have no choice in whether to develop a quantum-level description of computational processes, because even operating a classical computer at the quantum scale will require a thorough consideration of quantum mechanics.

However, the most compelling argument in favor of quantum computing comes from the scientific possibility afforded by qubits, that is, quantum bits. At no time since the beginning of classical digital computing in the 1930s, and never in classical communication theory, has there been a challenge to the basic unit of information—the bit. The qubit is qualitatively new; it is a chance to rework decades of theory and ideas using an entirely new building block for information. It is early yet, but the potential computational power afforded by qubits, and even more exotic types of "entangled" quantum bits, appears great.

An Overview of This Book

In the remainder of this book, we shall investigate the power and capabilities of computers based on quantum physics technologies. Chapter 2 explains how to achieve the key operations of a computer at the quantum scale and walks you through an example of quantum computation. Chapter 3 explores quantum complexity, the theory that enables us to estimate how much more efficiently a quantum computer can solve a problem than a classical computer. We look at Lov Grover's super-fast quantum algorithm for finding someone's name in a "quantum telephone directory" when all you know is his or her number. Although this sounds like a "toy" problem, Grover's algorithm can also be used to speed up the solution of many problems that can be couched as a search problem. Chapter 4 examines Peter Shor's quantum algorithm for breaking a top-secret code used by banks, government agencies, and people who conduct electronic transactions over the Internet. Chapter 5 discusses the concept of true randomness, its many facets and uses, and its connection to quantum computing. Chapter 6 looks at using quantum computing to establish a communication channel that is guaranteed to be secure, even if your adversary is a gifted mathematical genius with access to unlimited computing power. Chapter 7 describes a real scheme for teleporting a quantum state. In quantum teleportation you disassemble a quantum state at one location, transmit its parts through separate communication channels, and then reincarnate the quantum state at your desired destination. It sounds like science fiction, but it has already been implemented experimentally! Chapter 8 considers how we can correct quantum computations that have gone awry and describes how good we need to be at error correction before we can quantum-compute forever. Chapter 9 describes some ingenious schemes for building quantum computers and reports on the progress that has been made. In particular, we describe a prototypical, specialized quantum computer that has already been built and that exploits quantum effects in nuclear magnetic spectroscopy, the same technology used in medical imaging. Finally, Chapter 10 describes the idea of "computing without computing." In quantum computing, merely having a computer that *could* return the answer to some computation if that computer were run, is enough to obtain the answer even though the computer is *not* run!

Quantum computing is a field that is evolving rapidly. The progress over the last 5 years has been breathtaking. Ideas that were once ridiculed have now been demonstrated to work experimentally. The next chapter outlines the essential ideas of quantum computing and traces the steps needed to solve a particularly simple, but highly illuminating, computational problem, that of coming to a decision faster than is possible using classical logic alone.

Two

Quantum Computing

When you come to a fork in the road, take it.
—*Yogi Berra*

We are already witnessing the tremendous advantages of reducing text, pictures, sound, video, telephone calls, and even television signals to a gush of 0s and 1s, or "bits," the fundamental nuggets of classical information. Bits serve as the common currency of modern information processing. In the last chapter we saw how the trend in miniaturization of computer technology is leading to the idea of using individual quantum systems, such as electrons, atoms, or photons, to encode bits. However, the essence of quantum computing is not merely the use of *small* physical systems to encode bits but rather the realization that the very act of diminishing the dimensions of a bit means that the quantum bits, or "qubits," that result possess entirely new properties that can be harnessed to accomplish novel types of computation and communication.

One of these new properties is the fact that a qubit can be placed in a quantum state that is partly 0 and partly 1 simultaneously. As we shall see, it is ultimately this property that allows a quantum computer to perform several classical computations at once, in parallel, as in taking all the forks in the road simultaneously. Another new property is that we cannot make a perfect copy of a qubit in an unknown quantum state. This property can be exploited to guarantee the integrity of certain quantum communication channels, be-

Table 2.1
Implicit assumptions in the theory of classical computation.

Assumption	Classically	Quantum Mechanically
A bit always has a definite value.	True	False. A bit need not have a definite value until the moment after it is observed.
A bit can only be 0 or 1.	True	False. A bit can be in a superposition of 0 and 1 simultaneously.
A bit can be copied without affecting its value.	True	False. A qubit in an unknown state cannot be copied without necessarily disrupting its state.
A bit can be read without affecting its value.	True	False. Reading a qubit that is initially in a superposition will change the value of the qubit.
Reading one bit of the computer has no affect on another (unread) bit.	True	False. If the bit being read is entangled with another qubit, reading one qubit will affect the other.
To compute the result of a computation you must run the computer.	True	False. If you have a quantum computer that *could* perform the computation *if* it were run, then the answer can be obtained even though the computer is *not* run (see Chapter 10).

cause it prevents an eavesdropper from copying the qubits passing down the channel, measuring one, and relaying the other to the intended recipient. A third non-classical property of qubits it that when we try to read a qubit, we cannot do so without disturbing its quantum state. Thus classical bits and quantum bits behave in significantly different ways.

In this chapter, we develop the idea of *computing* with qubits. That is, we describe how to create quantum analogs of the basic components of a computer. In particular, we shall describe the idea of a quantum memory register that can hold collections of qubits just as a classical memory register holds collections of bits. We shall also describe how to cause the state of a quantum memory register to evolve in a way that can be interpreted as performing a useful computation. As we shall see, one of the principal advantages of quantum computers is that they can compute certain *joint* properties of all the possible answers to a particular computational problem in the time it takes a classical computer to find just one of the answers. This gives quantum computers the potential to be much faster than any classical computer, even a state-of-the-art supercomputer.

Tricks of the Trade

If you were to put a quantum computer under a microscope (a very powerful microscope), what phenomena might you see at work? To date, scientists have discovered how to harness only a handful of quantum phenomena in the service of computation. Nevertheless, these few phenomena are enough to suggest remarkable computing and communications devices. To date, the most useful phenomena have been found to be *superposition, interference, entanglement, non-determinism,* and *non-clonability.* At some point in the book we will examine each of these ideas in detail, but it will be helpful at the outset to give an overview of these phenomena.

The first phenomenon we shall describe is *superposition.* We saw this in Chapter 1 for the case of a single qubit. When you "measure," "read," or "observe" (we use these terms synonymously) a qubit it is always found to be in one of two states: 0 or 1. The Principle of Superposition says that if a quantum system can be measured to be in one of a number of states, then it can also exist in a blend, or superposition, of all of its observable states simultaneously. In the context of quantum computing, superposition means that an n-qubit memory register can exist in a superposition of all 2^n possible configurations of n classical bits. Superposition therefore allows a quantum computer to operate like a massively parallel computer, working on several (classical) bit string inputs at once. Unfortunately, it is impossible to observe these parallel computations individually, so we are severely limited in the information we can extract from the parallel computations.

The second phenomenon that arises in quantum computing is *interference.* Because the quantum computer can work on several classical inputs at once, we can cause those separate computations to interfere with, or influence, one another. In a sense, the computations going on in parallel can reinforce each other or cancel each other out, resulting in a net computational state that reveals a *collective* property of all the computations. This is the hallmark of quantum interference. You may have noticed such an interference effect when bright sunlight shines through net curtains. You see dark and bright fringes according to whether the light waves cancel or reinforce one another. The overall pattern that you see reveals collective information about the spacing of the holes in the net curtain.

The third phenomenon is *entanglement*. If two or more qubits are made to interact, they can emerge from the interaction in a definite *joint* quantum state that cannot be expressed in terms of a product of definite *individual* quantum states. Thus, the state of the component qubits can be fuzzy, even though the state of the collection of qubits as a whole is well defined. A consequence of entanglement is that a pair of entangled qubits retain a lingering, instantaneous influence on one another no matter how far apart they become and regardless of the nature of the intervening medium. If one of the qubits is measured, for example, then *at that instant,* the state of the qubit with which it is entangled becomes definite and, consequently, completely predictable, even though the two qubits might be in different galaxies. Such ghostly "non-local" effects, as they are called, are the principal source of the power of quantum computers over their classical counterparts. Indeed, non-locality is the *quintessential* quantum phenomenon. Whereas superposition and interference effects can also be seen in classical wave phenomena, such as ripples on the surface of a pond, non-locality has no counterpart in classical physics. It is purely a quantum phenomenon.

Our next phenomenon is *quantum non-determinism*. Non-determinism makes for the quantum equivalent of a fair coin toss: It is the inability to predict the quantum state into which a superposed state will collapse upon being measured. In the context of quantum computing, non-determinism means that if we prepare a qubit in the superposition $(1/\sqrt{2})(|0\rangle + |1\rangle)$ and then measure it, we cannot, either in practice or as a matter of principle, predict whether we will obtain a 0 or a 1 as the outcome of the measurement.

Finally, the last great idea of quantum computing is *non-clonability*. It turns out to be impossible to copy an unknown quantum state exactly. In other words, if you ask a friend to prepare a qubit in a superposed state $c_0|0\rangle + c_1|1\rangle$, without telling you what values he or she picked for c_0 and c_1, then it is impossible for you to make a perfect copy of the state that your friend prepared. This inability to copy unknown quantum states exactly is called quantum non-clonability. Non-clonability is intimately connected to the Heisenberg Uncertainty Principle, which states that it is impossible to measure certain pairs of quantities, such as the position and momentum of a particle, exactly simultaneously. However, if you could make a perfect quantum copy of the particle before you made any measurements, then you could use one copy to measure position and the other to measure momentum, and thereby skirt around the prohibitions of the Uncertainty Principle. Thus non-clonability and the Heisenberg Uncertainty Principle are two sides of the same coin. Even more baffling, if you could make a perfect copy of an unknown quantum state, you could send messages faster than the speed of light. But we'll postpone discussion of that trick until Chapter 7.

To recap, then, the key quantum phenomena on which quantum computing relies are superposition, interference, entanglement, nondeterminism, and non-clonability. Table 2.2 lists some of the applications of these phenomena, together with the chapter in which you will find a more detailed discussion of each.

There is little doubt that as quantum computing matures, more and more quantum phenomena will be enlisted into the service of computation.

Table 2.2
Some quantum phenomena and their applications.

Phenomenon	Example Application
Superposition	Quantum database search (Chapter 3) relies on superposition to load a quantum register with a state representing 2^n numbers in just n steps.
Interference	Quantum parallelism (Chapter 2) relies on interference to extract a joint property of all the solutions to a computational problem.
Entanglement	Quantum factoring (Chapter 4) relies on entanglement to create a repeating sequence of numbers whose period reveals the factors of a large integer. Quantum teleportation (Chapter 7) relies on entanglement to establish a non-local link between source and receiver of the state to be teleported.
Non-determinism	Quantum key distribution (Chapter 6) relies on non-determinism to guarantee that any eavesdropping will be detected.
Non-clonability	Quantum cryptography (Chapter 6) relies on non-clonability to guarantee security.
Non-locality	Quantum teleportation relies on non-locality and entanglement to disassemble and re-assemble the quantum state to be teleported.

Quantum Memory Registers

In order to be useful, any computer, even a quantum computer, must be capable of performing a certain minimal set of actions. In particular, it must be possible to program the computer, to input data, to run the computer, and to read off some final answer. In the quantum context, each of these actions has to be specified more carefully than is the case for classical machines. A good way to approach the challenge of designing a workable quantum computer is to start with the most rudimentary concept imaginable—that of a *quantum memory register.*

If a physical system is to act as a computer, it must possess some sort of memory register. In a classical computer, a memory register is used to hold any data—that is, bits—that are fed into the computer and to record any intermediate results that arise during the course of a long computation. A quantum memory register performs similar duties for quantum bits. Thus a quantum memory register can be thought of as a kind of quantum scratch pad that can be used to keep track of a computation.

Physically, a quantum memory register can be pictured as a string of qubits (see Figure 2.1).

You will recall from Chapter 1 that we can picture each qubit as an arrow contained in a sphere. If the arrow points vertical upward, the qubit is in state $|0\rangle$; if it points vertically downward, the qubit is in state $|1\rangle$; if the arrow is tilted horizontally to the left (see Figure 2.2), the qubit is in the superposition state

Figure 2.1   General state of a 2-qubit memory register.

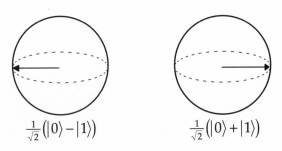

Figure 2.2   Two possible superposition states of a single qubit.

$(1/\sqrt{2})(|0\rangle - |1\rangle)$; if it is tilted horizontally to the right, the qubit is in superposition state $(1/\sqrt{2})(|0\rangle + |1\rangle)$.

When we "read" each qubit in an n-qubit memory register, it is always found to be in one of two distinct states. This means that a string of n qubits can be observed to be in one of $2 \times 2 \times \cdots \times 2 = 2^n$ (2 multiplied by itself n times) distinct states. These are precisely the binary (base-2) representations of the numbers from 0 to $2^n - 1$.

A base-2 (binary) number consisting of n bits can be converted to a base-10 ("normal") number as follows. Let b_j be the jth bit in the binary number counted from right to left and starting the count at $j = 0$. The binary number $b_{n-1}b_{n-2}\cdots b_1 b_0$ is just another way of writing the base-10 number $b_{n-1}2^{n-1} + \cdots + b_1 2^1 + b_0 2^0$. In other words, if the memory register consists of four qubits, the number zero is 0000, 1 is 0001, 2 is 0010, 3 is 0011, 4 is 0100, and so on, until 15 is 1111. Thus, in terms of the states in which we can measure the memory register to be, the classical and the quantum memory registers behave identically. You might think, therefore, that there is nothing to be gained by using a quantum memory register, because as soon as we look at the register, we are immediately thrown back into the world of classical information: zeroes and ones. What makes quantum computers so interesting, however, is what goes on when we're *not* looking.

Provided that we do not look at the quantum memory register, it can exist in a superposition state consisting of a blend of *all* the classically allowed configurations. For example, at any moment, a 4-bit classical register can be in one of 16 possible configurations: 0000, 0001, 0010, 0011, ..., 1111. A 4-qubit quantum register can be in any of these states too, but it can also be in a superposition state $|\psi\rangle = c_0 |0000\rangle + c_1 |0001\rangle + c_2 |0010\rangle + \cdots + c_{15} |1111\rangle$, where the numbers $c_0, c_1, c_2, \ldots, c_{15}$ are related to the proportions of each of the 16 possible classical bit string configurations to the net quantum state, and $|c_0|^2 + |c_1|^2 + \cdots + |c_{15}|^2 = 1$. Note that this superposition refers to the state of an *entire quantum register* of n qubits rather than to the single-qubit superposition we discussed in Chapter 1.

Superposition makes a quantum computer into a massively parallel computer. It allows a quantum computer to work on all possible classical inputs simultaneously in the time that it takes to work on just one of the inputs on a conventional computer. Unfortunately, there is no free lunch: Quantum mechanics severely limits our ability to access the information stored in this superposition. If we were to read the contents of the memory register that was in a superposition state, we would not see that superposition directly. Instead, as we read off each bit value one by one, the register would progressively collapse down to a single classical bit string. Thus, we would end up extracting only *one* of the answers that was in the original superposition.

If we were really clever, of course, we might be able to arrange for a quantum memory register to evolve in such a way that it would end up in a superposition of only acceptable solutions. In essence, this is exactly how some quantum computations work. Now that we have an idea how to store lots of information in a quantum memory register, our next task is to understand how to cause that memory register to evolve so that the natural physical evolution of the memory register leaves the register in a quantum state that, when measured, will reveal an answer to the desired computation.

The PREPARE–EVOLVE–MEASURE Cycle

Whereas a classical computer operates on a LOAD-RUN-READ cycle—that is, we load in the program and data, run the program, and read the answer—a quantum computer operates on an analogous PREPARE-EVOLVE-MEASURE cycle. The preparation step involves placing the quantum memory register in some initial state, typically with all its qubits in the $| 0 \rangle$ state. The evolution step involves stepping our quantum computer through a sequence of operations, which have the effect of transforming the initial state of the quantum memory register into a superposition of acceptable answer states. A subsequent measurement of this superposition will reveal one of these answers.

The state stored in our quantum memory register characterizes the instantaneous state of our computation, but does not tell us anything about how to cause that state to change, or evolve, in a manner that can be interpreted as the execution of some desired program. Achieving the desired evolution amounts to "programming" the quantum computer. In the context of a classical computer, either a program is a sequence of bits or, if the program is "hard-wired" into the device, it is simply an electronic circuit. If you own a digital watch, you are already carrying around a specialized "hard-wired" classical computer: one that simply counts the oscillations of a quartz crystal and displays the result. Likewise, it is easiest to think of a quantum computer as a specialized piece of hardware that has been carefully crafted to perform a single, but presumably important, quantum computation.

Back in 1926, long before quantum computers were ever imagined, physicist Erwin Schrödinger gave us the equation describing how a quantum computer evolves in time. This is because, as we have said before, a quantum computer is just a physical system. All isolated, non-relativistic quantum systems evolve in accordance with Schrödinger's equation. Therefore, our quantum computer must do so.

The Copenhagenist view of the evolution of a quantum memory register says that it evolves in two radically different ways, depending on whether it is being observed. If *unobserved,* the system will undergo a smooth, continuous evolution governed by Schrödinger's equation. However, if the quantum system *is* observed, it will appear to undergo a sudden, discontinuous, and unpredictable jump into one of the classical bit string configurations that can be written within the confines of the memory register.

Schrödinger's equation has a few remarkable features. First, it is a deterministic differential equation. This means that it can be used to predict the exact course of the future evolution of a quantum memory register or, indeed, the state of the register before this time. Thus Schrödinger's equation enables us to predict the future or past evolution of the memory register. This will become important later. It means that any operation that purports to be part of some grand quantum computation must be able to be run forwards or backwards in time; that is, the operation must be invertible. If it were not invertible, the operation could not possibly be described by Schrödinger's equation (which is always invertible), and so it could not be implemented using quantum physics. Second, Schrödinger's equation resembles the equations used to describe waves in classical physics. This should come as no surprise, because we saw in Chapter 1 that quantum systems can behave sometimes like waves and sometimes like particles. The waves in Schrödinger's equation are waves of probability amplitude.

If we solve Schrödinger's equation to determine what the state of the quantum memory register will be at an arbitrary time in the future, we find that the entire evolution can be expressed, very succinctly, as a particular *rotation* of the arrow representing the initial state of our quantum memory register. This rotation is smooth and continuous so long as we do not attempt to read any of the qubits in the memory register. The moment we do try to read a qubit, the smooth rotation ceases and the arrow is flung against one of the axes that corresponds to just one bit string configuration.

The result of any measurement of a quantum system that is in a particular quantum state is always one of a finite set of real numbers, the possible values of the observable being measured. If the system is in a special state called an eigenstate of the observable, then the outcome of the measurement will be the eigenvalue corresponding to this eigenstate. However, if the system is in a superposition of states, we can think if it as being composed of a weighted sum of the eigenstates such that the weights are complex numbers whose square moduli add up to 1. If a system whose state is described by a superposition is measured, the probability of finding it in a particular eigenstate is given by the square of the absolute value of the amplitude corresponding to that eigenstate in the superposition.

Thus, if the memory register of a quantum computer is in a superposition of possible states, then the result of a measurement of the state of the register will not be completely predictable, although we can predict the relative probability of each possible answer.

This relative probability is the probability of finding the *whole* memory in a particular configuration (a particular sequence of bits) when it is finally observed. Moreover, measurements of a subset of the qubits in the register project out the state of the whole register into a subset of eigenstates consistent with the answers obtained for the measured qubits.

Remarkably, the laws of quantum mechanics happen to be such that the multiple "parallel" computations that are affected by a quantum computer fed a superposition of inputs all proceed simultaneously to give rise to a superposition of outputs. If you try to measure the output state naively, you gain nothing over a classical computation. All you obtain is one of the output answers in the superposition. However, you can make certain measurements of joint properties of all the answers. In certain circumstances, this does provide a net win over classical computers. We'll have more to say about this at the end of the chapter.

Quantum Gates and Quantum Circuits

Quantum gates are analogous to classical logic gates, such as the AND and NOT gates, except that quantum gates must always have as many outputs as they have inputs. The latter requirement ensures that the output contains enough information so that the operation performed by the gate can be undone— "reversed" or "inverted"— without ambiguity. All isolated quantum systems have to obey this principle. It is a direct consequence of the mathematical form of Schrödinger's equation, which describes how quantum systems evolve. Because a quantum gate is just a special, isolated, quantum system, it too must operate in accordance with Schrödinger's equation and therefore must be invertible.

What does a quantum gate do to the state of the qubits on which it acts? The best way to think about this is to visualize the action of a 1-qubit gate on a single qubit. As we saw in Chapter 1, a single qubit can be pictured as an arrow extending from the center to the surface of a sphere. A 1-qubit gate provides mathematical rule for rotating this arrow to some new point on the surface. Different 1-qubit gates give rise to different rotations. Likewise, although it is impossible to visualize, the state of a pair of qubits can be pictured as a single arrow pointing in a four-dimensional space. A 2-qubit gate rotates this arrow such that it points in some new direction in this space. Thus the effect of a quantum gate on the qubits in a quantum memory register can be pictured as a rotation of the arrow that represents the state of that memory register.

To make this more clear, let's focus on the simplest possible quantum gate, the "Square-Root-of-NOT" gate ($\sqrt{\text{NOT}}$). This gate acts on a single qubit, so it has one input and one output. Two $\sqrt{\text{NOT}}$ gates, connected back to back, perform the NOT operation overall. The NOT operation, U_{NOT}, acting on the two observable states of a 1-qubit memory register is defined by

$$U_{\text{NOT}}|0\rangle = |1\rangle$$
$$U_{\text{NOT}}|1\rangle = |0\rangle$$

Thus the Square-root-of-NOT operation is defined by

$$U_{\sqrt{\text{NOT}}}U_{\sqrt{\text{NOT}}}|0\rangle = |1\rangle$$
$$U_{\sqrt{\text{NOT}}}U_{\sqrt{\text{NOT}}}|1\rangle = |0\rangle$$

If you think about this for a moment, you will realize that there is no way to do this if you restrict yourself to classical bits. That is, there is no way to define a set of mappings between 0 and 1 states such that two applications of the gate will succeed in inverting both 0 and 1.

However, there are quantum gates that manipulate single qubits that performs this feat. The key feature is that a quantum $\sqrt{\text{NOT}}$ gate is not restricted to mapping bits into bits. Instead it maps bits into superpositions of bits, and vice versa. For example, we could define a $\sqrt{\text{NOT}}$ gate as follows:

$$U_{\sqrt{\text{NOT}}}|0\rangle = \left(\frac{1}{2} + \frac{i}{2}\right)|0\rangle + \left(\frac{1}{2} - \frac{i}{2}\right)|1\rangle$$

$$U_{\sqrt{\text{NOT}}}|1\rangle = \left(\frac{1}{2} - \frac{i}{2}\right)|0\rangle + \left(\frac{1}{2} + \frac{i}{2}\right)|1\rangle$$

Thus, you see just how weird the quantum gates are in comparison to classical gates. Instead of manipulating good old zeroes and ones, our quantum gates manipulate complex weighted superpositions of $|0\rangle$ and $|1\rangle$. See Figure 2.3.

To make a quantum circuit that performs the NOT operation, built from two $\sqrt{\text{NOT}}$ gates, an input state representing the binary value 0 or the binary value 1 is transformed into a superposition of 0 and 1. This superposition is then fed into a second $\sqrt{\text{NOT}}$ gate, and the superposition is mapped into a gate corresponding to the opposite bit given in the input to the first gate. See Figure 2.4.

At the intermediate stage of the computation, after the action of just the first $\sqrt{\text{NOT}}$ gate, the state of the computation is a superposition of bits and, hence, is unlike any logic value used in conventional, classical computers.

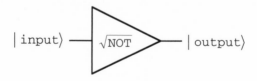

Figure 2.3 A quantum gate that acts on a single qubit.

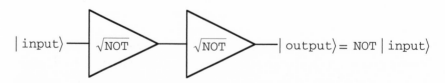

Figure 2.4 Two $\sqrt{\text{NOT}}$ gates connected back to back function as one NOT gate overall.

A 1-Qubit Rotation Gate

We can generalize the notion of the Square-Root-of-NOT gate to a 1-qubit gate that can apply an arbitrary rotation to the state of a single qubit. Three angles are needed to specify this rotation exactly. Because these angles can be any numbers whatsoever, there must be infinitely many 1-qubit gates.

An especially important 1-qubit gate is the so-called Walsh–Hadamard gate. This gate is defined by the following transformations:

$$U_{\text{WH}}|0\rangle = \frac{1}{\sqrt{2}}(|0\rangle + |1\rangle)$$

$$U_{\text{WH}}|1\rangle = \frac{1}{\sqrt{2}}(|0\rangle - |1\rangle)$$

This corresponds to a rotation of the state of the qubit to the halfway point between the $|0\rangle$ state and the $|1\rangle$ state. The rotation leaves the arrow, which represents the state of the qubit, pointing to the right if we start from the $|0\rangle$ state and pointing to the left if we start from the $|1\rangle$ state. Consequently, the Walsh–Hadamard gate can be used to create a superposition state out of a $|0\rangle$ state or a $|1\rangle$ state. However, note that the sign with which the $|1\rangle$ state appears in the superposition depends on whether we start with a $|0\rangle$ or a $|1\rangle$. Although this sign difference does not affect the *probability* of obtaining a 0 or 1 upon measuring either $(1/\sqrt{2})(|0\rangle + |1\rangle)$ or $(1/\sqrt{2})(|0\rangle - |1\rangle)$, it does have an important effect whenever we bring about quantum-mechanical interference between these states. Such interference effects are actually vital components of quantum computations.

The Walsh–Hadamard gate has another important application in quantum computing. If we initialize a quantum memory register so that each qubit in the register starts out in the $|0\rangle$ state and then apply the Walsh–Hadamard gate to each qubit independently, the net result places the entire n-qubit register in a superposition of all possible bit strings that an n-bit classical register can hold. Thus, using the Walsh–Hadamard gate, we can effectively enter 2^n bit strings into a quantum memory register using only n basic operations! For example, if we apply the Walsh–Hadamard gate to each of n qubits individually, we obtain the superposition of the 2^n numbers that can be represented in n bits.

$$\underbrace{U_{WH}|0\rangle \otimes U_{WH}|0\rangle \otimes \cdots \otimes U_{WH}|0\rangle}_{U_{WH} \text{ applied to } n \text{ qubits individually}}$$

$$= \frac{1}{\sqrt{2}}\big(|0\rangle + |1\rangle\big) \otimes \frac{1}{\sqrt{2}}\big(|0\rangle + |1\rangle\big) \otimes \cdots \otimes \frac{1}{\sqrt{2}}\big(|0\rangle + |1\rangle\big)$$

$$= \frac{1}{\sqrt{2^n}}\big(|00,\ldots,0\rangle + |00,\ldots,1\rangle + \cdots + |11,\ldots,1\rangle\big) \qquad \text{(in base 2 notation)}$$

$$= \frac{1}{\sqrt{2^n}}\big(|0\rangle + |1\rangle + |2\rangle + |3\rangle + \cdots + |2^n - 1\rangle\big) \qquad \text{(in base 10 notation)}$$

Thus, we can effectively load exponentially many (i.e., 2^n) numbers into a quantum computer using only polynomially many (i.e., n) basic gate operations. This is fortunate because, if it took exponentially many operations to load all the inputs into a quantum computer, then the gains in computational efficiency of quantum computing would be moot.

The Controlled-NOT Gate

There is a particular 2-qubit gate that is of paramount importance in quantum computing. It is called the Controlled-NOT gate, which we'll abbreviate as U_{CN}. The action of U_{CN} on the four observable states of a 2-qubit memory register—that is, 00, 01, 10 and 11—is as follows:

$$U_{CN}|00\rangle = |00\rangle$$
$$U_{CN}|01\rangle = |01\rangle$$
$$U_{CN}|10\rangle = |11\rangle$$
$$U_{CN}|11\rangle = |10\rangle$$

Can you see why U_{CN} is called the Controlled-NOT gate? The effect of U_{CN} is to flip (apply NOT to) the second qubit if the first qubit is 1 and to leave the second qubit alone otherwise. Note that this operation involves no measurements whatsoever; we do not need to measure qubits to bring about "controlled" operations.

Quantum circuits are usually drawn as sets of parallel quantum wires connecting various quantum gates. In a quantum circuit, Controlled-NOT appears as shown in Figure 2.5.

The standard symbol for the XOR (exclusive-OR) operation on elementary logic theory is $\oplus$. XOR takes two input bits and returns 1 if either, *but not both*, of its inputs is 1 and returns 0 otherwise. One of the outputs of the Controlled-NOT gate is written as $|x \oplus y\rangle$ because the effect of the Controlled-NOT gate is to perform an XOR operation on its two inputs, $|x\rangle$ and $|y\rangle$.

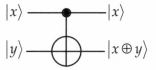

Figure 2.5   A Controlled-NOT gate as it would appear in a quantum circuit.

Universal Gates

The set of all 1-qubit rotations (which is an infinite set), together with the Controlled-NOT gate, is enough to achieve any imaginable quantum computation. That is, we can perform any quantum computation by connecting just 1-qubit rotation gates and Controlled-NOT gates.

A set of logic gates is said to be universal if any imaginable computation can be accomplished in a circuit comprising only gates of the type found in that set. Therefore, the set of all 1-qubit rotation gates and Controlled-NOT gates are universal for quantum computation.

In classical computation, it is well known that there are many universal sets of gates. For example, any classical computation can be accomplished using a circuit built out of AND and NOT gates. A similar result holds for quantum computation. Adriano Barenco and collaborators, at Oxford University, have shown that a certain 2-qubit gate is universal for quantum computation (Barenco, 1995). David Di Vincenzo of IBM arrived independently at a similar conclusion (DiVincenzo 1995a). Thus any feasible quantum computation can be accomplished by connecting just universal quantum gates. This means that, in principle, the designer of a quantum computer need only understand how to implement a few types of gates in order to implement any quantum computation. In practice, however, there may be more specialized quantum gates that would make it possible to build more compact circuits for specific computations.

Example of a Quantum Computation

Now we are ready to take a detailed look at a simple quantum computation. We are going to study what is called the Deutsch–Jozsa problem (Cleve et al. 1997, Ekert and Jozsa, 1998). This problem can be given a dramatic interpretation in the context of timely decision making. But it really boils down to computing a joint property of all of the outputs of a certain computation across all of its allowed inputs.

We suppose that we have discovered a guaranteed method for predicting whether a stock price will rise or fall tomorrow on the basis of whether a magical formula, f, returns the same or opposing answers on the pair of inputs 0 and 1. In other words, if $f(0) = f(1)$, then the stock price will certainly rise tomorrow, and if $f(0) \neq f(1)$, the stock price will certainly fall tomorrow. The trouble is that the formula f is so complicated that it takes a whole day to evaluate $f(0)$ and a whole day to evaluate $f(1)$. Thus it would takes our classical computer two full days to determine whether $f(0) = f(1)$. As a result, by the time we find out whether $f(0) = f(1)$, it is too late for that information to be useful. What are we to do, assuming, of course, that we are unable to buy a faster classical computer or a second classical computer? It turns out that if we have a quantum computer, we can indeed obtain our investment decision within the required time.

Each step in the computation that we are about to describe can be understood as a simple transformation of the state of a 2-qubit quantum memory register. Although the state transformations might appear daunting at first, if you follow through the operations in a step-by-step-fashion, you will soon see that they are quite straightforward. All you need to know is that -1 raised to the power 0 equals 1 and that -1 raised to the power 1 equals -1. In other words, $(-1)^0 = 1$ and $(-1)^1 = -1$.

It is very natural to think of a computer as something that transforms, or "maps," inputs into outputs. Mathematically, such a transformation is called a *function*. If you have studied elementary trigonometry, you have already encountered the sine, cosine, and tangent functions. Given an angle θ, the sine and cosine functions map this angle into $\sin(\theta)$ and $\cos(\theta)$, which are numbers between 1 and -1, and the tangent function maps the angle θ into $\tan(\theta)$, a number between minus infinity and plus infinity.

Mathematicians can dream up many other kinds of functions. For example, it is possible to define a function whose inputs are allowed to be only 0 or 1 and whose outputs are allowed to be only 0 or 1. The NOT function is an example. In this case, NOT(0) is 1 and NOT(1) is 0.

Let's imagine a problem in which we are given a magical function, like f, and are asked to decide whether the two outputs are the same—that is, $f(0) = f(1)$—or differ—that is, $f(0) \neq f(1)$. Using a classical computer, we would simply evaluate f on the input 0 and evaluate f on the input 1 and then compare the two results to decide whether $f(0)$ was indeed equal to $f(1)$. Thus, our classical computer requires a grand total of *two* function evaluations, and therefore takes two days of computer time to come up with the answer. However, using a quantum computer, we can decide the question in only one function evaluation, which means that we can reach our decision in 24 hours, in time to act on the advice of our magical formula.

The key to the quantum speedup lies in the fact that we are not asking what the values of the function $f(0)$ and $f(1)$ actually *are* but only whether they are the *same*. This is a subtle but important difference. It means that we are concerned with a question about the *joint* properties of both outputs rather than a question about the outputs directly, and quantum computers excel at answering questions about joint properties.

Here's how the quantum algorithm works. We want to decide whether our magical formula yields the same answer on both inputs, $f(0) = f(1)$, or yields different answers, $f(0) \neq f(1)$. Thus, we start by imagining a 1-qubit register in which we shall place the value $|x\rangle = |0\rangle$ or $|x\rangle = |1\rangle$. We then imagine designing a unitary operator that transforms $|x\rangle$ into $|f(x)\rangle$. Unfortunately, we see immediately that there is a problem. If $f(x)$ happens to return the same output for both inputs (which of course is precisely what we are hoping), then the process we envision for transforming the state that represents x into the state that represents $f(x)$ is not reversible and is not, therefore, implementable quantum–mechanically. This is because all isolated quantum systems must evolve in a manner that can be run backwards or forwards in time. However, if two different inputs yield the same output, there is no way that this process can be inverted unambiguously. The simplest way to fix this problem is to introduce an extra qubit that simply remembers the value of the input bit.

Now our problem is as follows: Given a 2-qubit register $|x\rangle|y\rangle$, we want to cause it to evolve into the state $|x\rangle|y \oplus f(x)\rangle$. Let's call the operation that does this U, so $U : |x\rangle|y\rangle \mapsto |x\rangle|y \oplus f(x)\rangle$. Here the operation $y \oplus f(x)$ is the exclusive-OR (XOR) operation defined by $0 \oplus 0 = 0$, $0 \oplus 1 = 1$, $1 \oplus 0 = 0$, and $1 \oplus 1 = 0$. Thus the transformation U that we want to bring about leaves the first qubit, x, unchanged, and performs a bit flip on the second qubit, y, if the function f applied to x is 1—that is, if $f(x) = 1$. Otherwise the operation leaves x and y unchanged.

If we can bring about the transformation U, we can determine whether $f(0) = f(1)$ or $f(0) \neq f(1)$ by running two tests: one in which we initialize the register to be $|x = 0\rangle|y = 0\rangle$, apply U, and measure the second qubit, and the other in which we initialize the register to be $|x = 1\rangle|y = 0\rangle$, apply U, and measure the second qubit. In both cases the input $|y\rangle = |0\rangle$, so the XOR operation simplifies to $|y \oplus f(x)\rangle = |f(x)\rangle$. Thus, if our two tests yield the same measurement outcome for the second qubit, we can be sure that $f(0) = f(1)$. Unfortunately, this procedure is no more efficient than the classical computation; it requires two evaluations of our magical function f. However, because we are now using quantum memory registers, we have the opportunity to start the register off in a superposition of states. Does this possibility offer any advantage?

Suppose that we initialize the register to be in the state $|x\rangle \otimes (1/\sqrt{2})(|0\rangle - |1\rangle)$—that is, we initialize the first qubit to be either $|0\rangle$ or $|1\rangle$ (it doesn't matter which, so we'll call it $|x\rangle$) and initialize the second qubit to be in the superposition state $(1/\sqrt{2})(|0\rangle - |1\rangle)$. The latter state can be created by initializing the second qubit to be in state $|1\rangle$ and then applying the Walsh–Hadamard operation, W, to this state.

Now let's see what happens when we apply our transformation U to this state. Remember that U brings about the transformation $U : |x\rangle|y\rangle \mapsto |x\rangle|y \oplus f(x)\rangle$, so starting with $|y\rangle$ in the superposition state $(1/\sqrt{2})(|0\rangle - |1\rangle)$, we get the transformation $|x\rangle \otimes (1/\sqrt{2}) \times (|0\rangle - |1\rangle) \mapsto |x\rangle \otimes (1/\sqrt{2})(|0 \oplus f(x)\rangle - |1 \oplus f(x)\rangle)$. However, because our magical function f always returns either 0 or 1, we can reinterpret this state. Consider each possibility: If $f(x) = 0$, the state of the 2-qubit register is $|x\rangle \otimes (1/\sqrt{2})(|0\rangle - |1\rangle)$. Conversely, if $f(x) = 1$, the state of the 2-qubit register is $|x\rangle \otimes (1/\sqrt{2})(|1\rangle - |0\rangle)$. Thus (watch out, here comes the crucial step) regardless of whether $f(x) = 0$ or $f(x) = 1$, we can re-express the state $|x\rangle \otimes (1/\sqrt{2})(|0 \oplus f(x)\rangle - |1 \oplus f(x)\rangle)$ as entirely equivalent to the state $|x\rangle \otimes (1/\sqrt{2})(-1)^{f(x)}(|0\rangle - |1\rangle)$, which, if you stare at it a little, you will recognize as the same as the state we started with, except that it now contains a phase factor, $(-1)^{f(x)}$, that depends on the magical function $f(x)$.

Why is this important? Well, it means that we have found a way of encoding the value of the function $f(x)$ in a phase factor. The next thing to do is figure out how to work on both inputs—that is, both values of x—simultaneously.

We can perform the operation U, described above, on both inputs simultaneously by allowing the state $|x\rangle$ to be a superposition of both $|0\rangle$ and $|1\rangle$—that is, by initializing the quantum state of the first qubit to be $|x\rangle = (1/\sqrt{2})(|0\rangle + |1\rangle)$. If we follow through the logic,

$$\frac{1}{\sqrt{2}}(|\,0\rangle+|\,1\rangle) \otimes \frac{1}{\sqrt{2}}(|\,0\rangle-|\,1\rangle)$$

$$\rightarrow \frac{1}{\sqrt{2}} \left((-1)^{f(0)}\,|\,0\rangle + (-1)^{f(1)}\,|\,1\rangle\right) \otimes \frac{1}{\sqrt{2}}(|\,0\rangle-|\,1\rangle)$$

The next thing to think about is how to exploit the fact that the values of f are encoded in a phase factor. Remember that the function f takes a 1-bit input and returns a 1-bit output. A bit can have one of two values, so this means that there are precisely four different ways in which f could be defined. Specifically, the function f may be $\{f(0) = 0, f(1) = 0\}$, $\{f(0) = 0, f(1) = 1\}$, $\{f(0) = 1, f(1) = 0\}$ or $\{f(0) = 1, f(1) = 1\}$. Let's call these cases A, B, C, and D, respectively. In cases A and D, the function f returns the *same* output for different inputs, so the $|0\rangle$ and $|1\rangle$ components of the first qubit are multiplied by the same phase factor, that is, $(-1)^{f(0)} = (-1)^{f(1)}$. Conversely, in cases B and C, wherein the function f returns different outputs for different inputs, the $|0\rangle$ and $|1\rangle$ components of the first qubit have opposing phase factors, that is, $(-1)^{f(0)} \neq (-1)^{f(1)}$. The final step of the quantum algorithm is to bring about quantum mechanical interference between the phase factors by applying an operation, say V, that imposes the transformation

$$\frac{1}{\sqrt{2}} \left[(-1)^{f(0)}|0\rangle + (-1)^{f(1)}|1\rangle\right] \otimes \frac{1}{\sqrt{2}}(|0\rangle - |1\rangle)$$

$$\rightarrow \frac{1}{\sqrt{2}} \left[\left((-1)^{f(0)} + (-1)^{f(1)}\right)|0\rangle + \left((-1)^{f(0)} - (-1)^{f(1)}\right)|1\rangle\right]$$

$$\otimes \frac{1}{\sqrt{2}}(|0\rangle - |1\rangle)$$

Having applied the operation V, we measure the state of the first qubit. This qubit will be found, with certainty, to be in the state $|0\rangle$ if $f(0) = f(1)$, regardless of whether $f(0) = f(1) = 0$ or $f(0) = f(1) = 1$. Conversely, the first qubit will be found, with certainty, to be in state $|1\rangle$ if $f(0) \neq f(1)$. Thus, the question of whether the magical function f returns the same output on different inputs— that is, $f(0) = f(1)$—or opposing outputs on different inputs has been answered using only *one* evaluation of the function f. Therefore, using a quantum computer and our magical function f, we can determine, *in time to make a handsome profit*, whether or not to invest in the stock.

Table 2.3
Solving Deutsch's Problem on a quantum computer.

---

1. Initialize the 2-qubit register in the state $|0\rangle|1\rangle$.

2. Apply the Walsh–Hadamard operation, W, to each qubit.

$$|0\rangle|1\rangle \rightarrow \frac{1}{\sqrt{2}}(|0\rangle + |1\rangle) \otimes \frac{1}{\sqrt{2}}(|0\rangle - |1\rangle)$$

3. Apply the operation U (which requires f to be evaluated once only).

$$\frac{1}{\sqrt{2}}(|0\rangle + |1\rangle) \otimes \frac{1}{\sqrt{2}}(|0\rangle - |1\rangle)$$
$$\rightarrow \frac{1}{\sqrt{2}}\left((-1)^{f(0)}|0\rangle + (-1)^{f(1)}|1\rangle\right) \otimes \frac{1}{\sqrt{2}}(|0\rangle - |1\rangle)$$

4. Apply the operation V (which does not require f to be evaluated).

$$\frac{1}{\sqrt{2}}\left[(-1)^{f(0)}|0\rangle + (-1)^{f(1)}|1\rangle\right] \otimes \frac{1}{\sqrt{2}}(|0\rangle - |1\rangle)$$
$$\rightarrow \frac{1}{\sqrt{2}}\left[\left((-1)^{f(0)} + (-1)^{f(1)}\right)|0\rangle + \left((-1)^{f(0)} - (-1)^{f(1)}\right)|1\rangle\right]$$
$$\otimes \frac{1}{\sqrt{2}}(|0\rangle - |1\rangle)$$

5. Measure the bit value in the first qubit. If it is 0, $f(0) = f(1)$. If it is 1, $f(0) \neq f(1)$.

At first glance, you might find this quantum computation rather confusing. It definitely requires a complete rethinking of computational logic. However, if you follow the steps carefully, you'll see that each is really quite simple. We have gathered them all together in Table 2.3 to help you grasp the overall plan.

In essence, the quantum algorithm exploits superposition and interference to extract a joint property of both function values, $f(0)$ and $f(1)$, without having to reveal either function value explicitly.

Summary

In this chapter we have explored the analogy between the LOAD-RUN-READ cycle of a classical computer and the PREPARE-EVOLVE-MEASURE cycle of a quantum computer. The most important idea is that just as a single qubit can be placed in a superposition state, so too can an entire quantum memory register consisting of n qubits. When we do so, however, it is possible to create special quantum states of the register that have no counterpart in the case of a single qubit. In particular, if two (or more) qubits interact, they invariably entangle with one another in the sense that their joint quantum state is definite even though their individual quantum states are indefinite. The simplest example is the 2-qubit state $(1/\sqrt{2})(|00\rangle + |11\rangle)$. Here if one of the qubits is measured and found to be in state $|1\rangle$, you can be certain, without even making the measurement, that the other qubit will also be in state $|1\rangle$. If such a pair of entangled qubits are separated, then a measurement made on one qubit can have an instantaneous effect on the other qubit, regardless of their distance apart and regardless of the material that lies between them.

The principal advantage of a quantum computer over a classical computer is that it can use a technique called quantum parallelism to compute certain *joint* properties of several superposed computations—several answers to different classical computations—in the time it takes a classical computer to find just one of the answers. Moreover, the quantum computer can do this without having to reveal the answer to any one of those computations individually. This gives a quantum computer the potential to be vastly more efficient than a classical computer at certain computational tasks. Just how much more efficient? That is the topic of the next chapter.

Three

━━◆◆◇◆◆━━

What Can Computers Do?

Computers are useless. They only give answers.
—Pablo Picasso

In Chapter 1, we discussed the trend toward miniaturization that is luring the computer industry into the unpredictable realm of quantum mechanics. In Chapter 2, we described how a computer that operates quantum–mechanically can harness exotic phenomena, such as entanglement and non-clonability, that have no parallels in the everyday world around us. The question is whether such phenomena confer an advantage. Do they make the capabilities of a quantum computer surpass those of a classical computer? This is an important question because it will require a massive financial investment to create quantum computers. We have to be able to determine whether the effort and expense will be justified.

To address this issue, we need to define *surpass* more precisely. First, there is the question of computational *complexity:* Can a quantum computer perform the same tasks as a classical computer in significantly fewer steps? Second, there is the question of *computability:* Can a quantum computer *do* computations that a classical computer cannot? And finally, there is the question of *universality:* Can a specialized quantum computer simulate any other quantum computer, or classical computer, efficiently? A difference between the capabilities of a quantum computer and those of a classical computer on any one of these criteria would be significant.

But how can we possibly characterize the capabilities of a computer? If you ask a young child what computers can do, you may be told, "They let me draw pictures and play games." Ask a teenager and you may hear, "They let me surf the Web and go online with my friends." Ask an adult and you might discover, "They're great for word processing and keeping track of my finances." What is remarkable is that the toddler, the teenager, and the parent might all be talking about the same machine! It seems that by running the appropriate software, we can make the computer perform almost any task.

But what exactly are the capabilities of computers? In particular, what problems can computers solve in principle and what problems can they solve in practice, that is, efficiently? Are there any problems that computers will never solve, no matter how powerful they become? Does it matter whether computers are implemented as gears, vacuum tubes, transistors, or integrated circuits? These questions can be answered by developing a mathematical model (or theory) of computation and exploring its ramifications.

It might seem preposterous that *any* theory could purport to circumscribe the capabilities of computers. Each year new computers appear that are significantly faster than their predecessors, and thousands more programs become available. People upgrade their computers precisely because they believe their new machines can do more than their old ones. If a theory of computation is based on particular machines and particular software, then it will quickly become outdated. This point was understood by the 1930s theorists Alan Turing, Alonso Church, Kurt Gödel, and Emil Post. They independently developed *mathematical* models of the computing process that were intended to be free of any assumptions about how computers were actually implemented.

Superficially, these models were quite different from one another. Gödel identified the tasks that a computer can perform with a class of mathematical functions that refer to themselves. For example, given a sequence of inputs, 1, 2, 3, 4, 5, 6, . . ., the Fibonnaci function generates the corresponding sequence of outputs 1, 1, 2, 3, 5, 8, . . ., in which successive outputs are the sum of the two previous outputs. The Fibonacci function can therefore be defined in terms of itself according to fibonacci(i) = fibonacci($i - 1$) + fibonnaci($i - 2$), provided that the self-reference bottoms out at fibonacci(1) = fibonacci(2) = 1. A second attempt at characterizing what computers can do was made by Church. He equated the tasks that a computer can perform with the " λ-definable functions" (which you have encountered if you have ever used the LISP programming language). Finally, Alan Turing identified the tasks a computer can perform with the class of functions computable by a hypothetical computing device which came to be called a Turing machine in his honor.

It turns out, however, that these competing models of "computation" are equivalent. This came as something of a surprise, because there was no reason a priori to expect their equivalence. Moreover, any one of the models alone might be open to the criticism that it provided an incomplete account of computation. But the fact that three radically different views of computation all turned out to be equivalent was a clear indication that the most important aspects of computation had been characterized correctly.

Unfortunately, we now know that although these models were intended to be mathematical abstractions of computation that were free of physical assumptions, they do, in fact, harbor rather subtle physical assumptions. These assumptions appear to be perfectly valid in the world we see around us, but they cease to be valid on sufficiently small scales. To appreciate at what point the classical models break down, it will be useful to take a look at the context in which the models arose.

The Turing Machine

The most influential model of computation was devised by Alan Turing in 1936 (Turing, 1937).[1] A Turing machine (TM) is an idealized mathematical model of a computer that can be used to understand the limits of what computers can do (Hopcroft, 1984). It is not meant to be a practical design for any actual machine but is rather a simplified abstraction that nevertheless captures the *essential* features of any real computer. A Turing machine's usefulness stems from its being simple enough to allow mathematicians to prove theorems about its computational capabilities, and yet complex enough to accommodate any actual classical digital computer, no matter how it is implemented in the physical world.

Turing's idea for such a machine grew out of an attempt to answer a question posed by David Hilbert, an eminent German mathematician. In 1900, at the International Congress of Mathematics held in Paris, Hilbert gave an address concerning what he believed to be the 23 most challenging mathematical problems of his day.

---

[1] Such are the vagaries of publication!

The last problem on his list asked whether there was a mechanical procedure by which the truth or falsity of any mathematical conjecture or question could be decided. In German, the word for "decision" is *Entscheidung,* so Hilbert's 23rd problem became known as the *Entscheidungsproblem.*

Hilbert's motivation for asking this question arose from the trend toward abstraction in mathematics. Throughout the nineteenth century, mathematics was largely a practical matter, concerned with making statements about real-world objects. In the late 1800s, mathematicians began to invent, and then reason about, imaginary objects to which they ascribed properties that were not necessarily compatible with "common sense." The truth or falsity of statements made about such imaginary objects could not be determined by appealing to the real world. In an attempt to put mathematical reasoning on secure logical foundations, Hilbert advocated a "formalist" approach to proofs. To a formalist, symbols cease to have any meaning other than that implied by their relationships to one another. No inference is permitted unless there is an explicit rule that sanctions it, and no information about the meaning of any symbol enters into a proof from outside the proof. Thus the very philosophy of mathematics that Hilbert advocated seemed very machine-like, and hence Hilbert proposed the *Entscheidungsproblem.*

Turing heard about Hilbert's *Entscheidungsproblem* during a course of lectures he attended at Cambridge University. Although Hilbert probably meant "mechanical" figuratively, Turing interpreted it literally. Turing wondered whether a *machine* could exist that would be able to decide the truth or falsity of any mathematical proposition. Thus Turing realized that in order to address the *Entscheidungsproblem,* he needed to model the process that a mathematician goes through when attempting to prove some mathematical conjecture.

Mathematical reasoning is an enigmatic activity. We do not really know what goes on inside a mathematician's head, but we can examine the result of the process in the form of the notes the mathematician creates while developing a proof. Mathematical reasoning consists of combining axioms (statements taken to be true without proof) with an ever-evolving set of conclusions, to infer new conclusions.

Turing abstracted the process followed by the mathematician into four principal ingredients: a set of transformation rules that allowed one mathematical statement to be transformed into another, a method for recording each step in the proof, an ability to go back and forth over the proof to combine earlier inferences with later ones, and a mechanism for deciding which rule to apply at any given moment. This is the essence of the proof process—at least the visible part of it.

Next Turing sought to simplify these steps in such a way that a machine could be made to imitate them. Mathematical statements are built up out of a mixture of ordinary letters, numbers, parentheses, operators (such as plus, $+$, and times, $\times$), and special mathematical symbols (such as $\forall, \exists, \neg, \wedge, \vee, \rightarrow, \leftrightarrow, \therefore,$ and $\because$). Turing realized that the symbols themselves were of no particular significance. All that mattered was that they were used consistently and that their number was finite. Moreover, once you know you are dealing with a finite alphabet, you can place each symbol in one-to-one correspondence with a unique pattern of any two symbols (such as 0 and 1). Hence Turing realized that rather than dealing with a rich array of esoteric symbols, a machine only needed to be able to read and write two kinds of symbols—say 0 and 1—with blank spaces or some other convention being used to identify the boundaries between the distinct symbols.

Similarly, the fact that the scratch pad on which the mathematician writes intermediate results is two-dimensional is of no particular importance. We can imagine attaching the beginning of one line of a proof to the end of the previous line, making one long continuous strip of paper. For simplicity, then, Turing assumed that the proof could be written out on a long strip of paper or a "tape." Moreover, rather than allowing free-form handwriting, it would clearly be easier for a machine to deal with a tape marked off into a sequence of identical cells, where only one symbol was permitted to be written inside each cell or the cell was left blank.

Finally, the process of the mathematician going back and forth over previous conclusions in order to draw new ones could be captured by imagining that there is a "read/write head" going back and forth along the tape.

When a mathematician views an earlier result, it is usually in some context. A mathematician might read a set of symbols and write something but later come back to read those same symbols again and write something else. Thus the context in which a set of symbols is read can affect the subsequent actions. Turing captured this idea by defining the "head" of his Turing machine to be in certain "states" corresponding to particular contexts. The combination of the symbol being read under the head and the state of the machine determined what symbol to write on the tape, which direction to move the head, and which state to enter next.

This crude model of the proof process turned out to be curiously powerful. No matter what embellishments people dreamed up, Turing could always argue that they were merely refinements to some existing part of the model rather than being fundamentally new features. Consequently, the Turing machine model indeed embodied the essence of the proof process.

Soon, by putting the aforementioned mechanistic analogs of human behavior into mathematical form, Turing arrived at the idea of what became known as a "deterministic Turing machine."

Deterministic Turing Machines

A deterministic Turing machine (DTM) is illustrated in Figure 3.1. Its components are inspired by Turing's abstract view of mathematical reasoning.

A deterministic Turing machine consists of an infinitely long tape that is marked off into a sequence of cells, on which may be written a 0 or a 1, and a read/write head that can move back and forth along the tape, scanning the contents of each cell. The head can exist in one of a finite set of internal "states" and contains a set of instructions (constituting the "program") that specifies, given the current internal state, how the state must change given the bit (the binary digit 0 or 1) currently being read under the head, whether that bit should be changed, and in which direction the head should then be advanced.

The tape is initially set up in some standardized state, such as all cells containing 0 except for a few that hold the program and any initial data. Thereafter, the tape serves as the scratch pad on which all intermediate results and the final answer (if any) are written.

Despite its simplicity, the Turing machine model has proved remarkably durable. In the 60-odd years since its inception, computer technology has advanced considerably. Nevertheless, the Turing machine model remains as applicable today as it was back in 1936. Al-

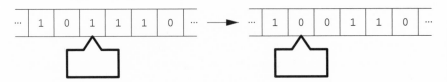

Figure 3.1 A deterministic Turing machine.

though we are apt to think of multimillion-dollar supercomputers as being more powerful than humble desktop machines, the Turing machine model proves otherwise. There is not a single computation that a supercomputer can perform that a personal computer cannot also perform, given enough time and memory capacity. In the strict theoretical sense, they are equivalent. Thus, the Turing machine is the foundation upon which much of current computer science rests. It has enabled computer scientists to prove many theorems that circumscribe the capabilities of computing machinery.

More recently, however, a new idea has emerged that adds a slight twist to the deterministic Turing machine. To understand this new twist, it is helpful to think of a computational problem as having a large number of potential solutions but only one actual solution. For example, suppose a friend picks a card from a deck of 52 cards and asks you to guess which one she picked. This problem is a search problem that has 52 potential solutions but only one actual solution. You could program a deterministic Turing machine to search through the deck of cards and ask, "Is this the card? Is this the card? Is this the card?" and so on. When a deterministic Turing machine reaches the state that corresponds to the actual solution, it halts and reports the answer. Unfortunately, deterministic Turing machines, which follow rigid predefined rules, are susceptible to systematic bias in the computational process that causes the Turing machine to take a very long time to solve certain problems. These are the problems for which the particular set of deterministic rules happen to make the Turing machine examine nearly all the potential solutions before discovering the one actual solution. As a consequence, if your friend knew the rules your deterministic card-picking Turing machine was to follow, she could choose the card as the last one the machine would examine. To avoid such pitfalls, a new type of Turing machine was invented that employs randomness. It is called a probabilistic, or non-deterministic, Turing machine.

Probabilistic Turing Machines

An alternative model of classical computation equips a deterministic Turing machine with the ability to make a random choice, such as flipping a coin. The result is a probabilistic Turing machine. Surprisingly, many problems that take a long time to solve on a deterministic Turing machine can often be solved very quickly on a probabilistic Turing machine (PTM).

In the probabilistic model of computation, there are often trade-offs between the time it takes to return an answer to a computation and the quality of the answer returned. For example, suppose you want to plan an around-the-world trip on which you visit 100 cities, but you want to minimize the distance you have to travel between cities and you want to visit each city only once. The problem of computing the optimal (shortest-path) route for your trip is extremely demanding computationally. However, if you are prepared to accept a route that is only guaranteed to be close to the optimal route, and that might in fact be the optimal route, then this problem is very easy to solve computationally. Here, then, is an example of a problem that is hard to solve if you want the exact answer but easy to solve if you are content with merely a good answer.

An alternative tradeoff, if you *require* a correct answer, is to allow uncertainty in the length of time the probabilistic algorithm must run before it returns an answer. Consequently, a new issue enters the computational theory: the correctness of an answer and its relationship to the running time of an algorithm.

Whereas a deterministic Turing machine, in a certain state, reading a certain symbol, has precisely one successor stage available to it, the probabilistic Turing machine has multiple legitimate successor states available (see Figure 3.2). The choice of which state to explore is determined by the outcome of a random choice (possibly with a bias in favor of some states over others). In all other respects, the PTM is just like a DTM.

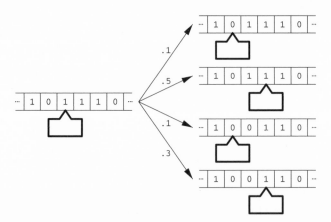

Figure 3.2 In a probabilistic classical Turing machine, there are multiple possible successor states, only one of which is actually selected. Unselected paths are terminated (×). The probabilities of transitioning between various states are shown. Note that the sum of the probabilities of all the paths emanating from a state is 1.

Despite the superficial difference between PTMs and DTMs, computer scientists have proved that anything computable by a probabilistic Turing machine can also be computed by a deterministic Turing machine, although in such cases the probabilistic machine is often more efficient (Gill, 1977). The basic reason for the success, of the probabilistic approach is that a probabilistic algorithm can be thought of as alternating among a collection of deterministic algorithms. Whereas it is fairly easy to design a problem so that it will mislead a particular deterministic algorithm, it is much harder to do so for a probabilistic algorithm, because it keeps on changing its "identity." Indeed, the latest algorithms for solving hard computational problems now interleave deterministic with probabilistic steps. The exact proportions in which these strategies are invoked can have a huge impact on the overall efficiency of problem solving.

Notwithstanding these successes, in the early 1980s a few maverick scientists began to question the correctness of the classical models of computation. The deterministic Turing machine and probabilistic Turing machine models are certainly fine as *mathematical* abstractions, but are they consistent with known *physics?* This question was irrelevant in Turing's era, because computers operated at a scale well above that of quantum systems. As miniaturization progresses, however, it is reasonable—in fact, necessary—to reconsider the foundations of computer science in the light of our improved understanding of the microscopic world. We now know that the Turing machine model contains fatal flaws. In spite of Turing's best efforts, some remnants of classical physics, such as the assumption that a bit must be either a 0 *or* a 1, crept into the Turing machine model (for a complete list of the faulty assumptions see Table 2.1). The obvious advances in technology (such as more memory, more instructions per second, and greater energy efficiency) have all been quantitative in nature. The underlying foundations of computer science have not changed. Similarly, although they have had a huge social impact, apparent revolutions such as the explosion of the Internet have merely provided new conduits for information to be exchanged. They have not altered the fundamental capabilities of computers in any way. As computers become smaller, however, eventually their behavior *must* be described in terms of the physics appropriate for small scales—that is, quantum physics.

The introduction of quantum considerations turn out to have profound implications for the foundations of computer science and information theory. Decades of old theory must now be taken from the library shelves, dusted off, and checked for an implicit reliance on classical physics and classical bits. By exploiting entirely new kinds of physical phenomena, such as superposition, interference, entanglement, non-determinism and non-clonability, we can suddenly catch a glimpse of a new theoretical landscape before us. This shift from classical to quantum is a *qualitative* change, not merely a *quantitative* change like the trends we saw in Chapter 1. It is something entirely new.

We can consider the impact of quantum mechanics on the *foundations* of computer science rather clearly if we think about "quantumizing" the abstract Turing model of computation. The net result of this endeavor is a model of a quantum Turing machine.

Quantum Turing Machines

The first quantum-mechanical description of a Turing machine was given by Paul Benioff (Benioff, 1980). He was building on earlier work carried out by Charles Bennett, who had shown that a reversible Turing machine was a theoretical possibility (Bennett, 1973).

A reversible Turing machine is a special version of a deterministic Turing machine that never throws any information away. This is important because physicists had shown that, in principle, *all* of the energy expended in performing a computation can be recovered, provided that the computer does not throw any information away. The notion of "not throwing information away" means that the output from each step of the machine must contain enough information so that the step can be undone without ambiguity. This requirement precludes the use of certain conventional logic gates, such as the AND gate, inside any reversible computer. The AND operation x AND y combines two bits x and y (usually thought of as "truth" values; True $= 1$ and False $= 0$) and returns their logical conjunction: 0 AND $0 = 0$, 0 AND $1 = 0$, 1 AND $0 = 0$, 1 AND $1 = 1$. Note that if the output from an AND gate is 1, we can infer, unambiguously, that the input was 11, that is, two ones. But if the output is 0, it is impossible to tell whether the input was 00, 01, or 10. Thus, if we think of a reversible Turing machine as a dynamical system, then knowledge of its state at any one moment would enable us to predict its state at all future and all past times. No information was ever lost, and the entire computation could be run forwards or backwards.

This fact struck a chord with Benioff, for he realized that any isolated quantum system had a dynamical evolution that was reversible in exactly this sense. Thus, it ought to be possible to devise a quantum system whose evolution over time mimicked the actions of a classical reversible Turing machine. This is exactly what Benioff did.

Unfortunately, Benioff's hypothetical machine is not quite a true quantum computer. Although between computational steps the machine exists in an intrinsically quantum state (in fact, a "superposition,"a blend of classical bit string configurations simultaneously), at the end of each step the "tape" of the machine is measured, projecting it back in one of its classical states: a sequence of bits. Thus Benioff's design can do no more than a classical reversible Turing machine.

The possibility that quantum-mechanical effects might offer something genuinely new was first hinted at by Richard Feynman of the California Institute of Technology (Caltech) in 1982 (Feynman, 1982), when he showed that no classical Turing machine could simulate certain quantum phenomena without incurring an unacceptable large slowdown but that a "universal quantum simulator" could do so. Unfortunately, Feynman did not provide a design for such a simulator, so his idea had little immediate impact.

The key step in making it possible to study the computational power of quantum computers came in 1985, when David Deutsch of Oxford University described the first true quantum Turing machine (QTM) (Deutsch, 1985). A QTM is a Turing machine whose read, write, and shift operations are accomplished by quantum-mechanical interactions and whose "tape" can exist in states that are highly nonclassical. In particular, whereas a conventional classical Turing machine can encode only a 0, 1, or blank in each cell of the tape, the QTM can exist in a blend, or "superposition," of 0 and 1 simultaneously corresponding to several different computations performed at once. Thus, the QTM has the potential for encoding many inputs to a problem simultaneously on the same tape and performing a calculation on all the inputs in the time it takes to do just one of the calculations classically. This results in a superposition of all the classical results, and with the appropriate measurement, information about certain joint properties of all these classical results can be extracted. This technique is called "quantum parallelism." We saw an example of quantum parallelism when we solved Deutsch's problem in Chapter 2.

Moreover, the superposition state representing the tape of the QTM can correspond to an entanglement of several classical bit string configurations. Entanglement means that the quantum state of the entire tape is well defined but the state of the individual qubits is not. For example, a 3-qubit tape in the state $(1/\sqrt{2})(|010\rangle + |101\rangle)$ represents an entanglement of the two configurations 010 and 101. It is entangled in the sense that if we were to measure any one of these qubits, the quantum state of the other two qubits would become definite instantaneously. Thus, if we read out the bit values from a part of the tape of the QTM when it is in an entangled state, our actions will have a side effect on the state of the other (unmeasured) qubits. In fact, the existence of such "entangled" qubits is the fundamental reason why QTMs are different from classical deterministic and probabilistic TMs.

As we saw in Chapter 1, each qubit in a QTM can be visualized as a small arrow contained in a sphere. "Straight up" represents the (classical) binary value 0, and "straight down" represents the (classical) binary value 1. When the arrow is at any other orientation, the angle the arrow makes with the horizontal axis is a measure of the ratio of 0-ness to 1-ness in the qubit. Likewise, the angle through which the arrow is rotated about the vertical axis is a measure of the "phase." Thus, by drawing qubits as arrows contained in spheres, we can depict Deutsch's quantum Turing machine as shown in Figure 3.3.

Quantum Turing machines (QTMs) are best thought of as quantum-mechanical generalizations of probabilistic Turing machines (PTMs). In a PTM, if we initialize the tape in some starting configuration and run the machine without inspecting its state for some time t, then its final state will be uncertain and can be described only by using a probability distribution over all the possible states accessible in time t.

Likewise, in a QTM, if we start the machine off in some initial configuration and allow it to evolve for a time t, then its state at time

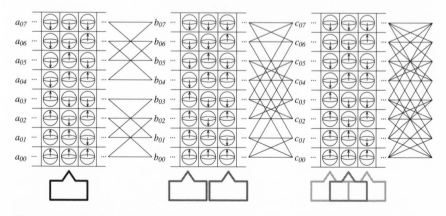

Figure 3.3 In the quantum Turing machine, each cell on the tape can hold a qubit, whose state is represented as an arrow contained in a sphere. *All* paths are pursued simultaneously.

t is described by a superposition of all states reachable in time t. The key difference is that in a classical PTM, only one particular computational trajectory is followed, but in the QTM all computational trajectories are followed and the resulting superposition is the sum over all possible trajectories achievable in time t. Moreover, in general, in the QTM the qubits will be entangled with one another. For example, suppose we focus on the state of just three of the qubits on the tape of the QTM. Let's call them qubits A, B, and C. At some point in the computation suppose that these qubits can only be either 000 or 111, and no other combinations are reachable from the starting state and transformation rules of the QTM. In such a situation, the state of the QTM might have form

$$\frac{1}{2}\left|\cdots \underbrace{0 \quad 0 \quad 0}_{\text{qubits } A, B \text{ and } C} \cdots\right\rangle + \frac{\sqrt{3}}{2}\left|\cdots \underbrace{1 \quad 1 \quad 1}_{\text{qubits } A, B \text{ and } C} \cdots\right\rangle.$$

At this moment, if we measure the bit-values of these three qubits, they are more likely to be found to be 111 than 000. However, if they are not measured, that is, if we do not look at the tape of the QTM, the bit-values of A, B, and C would remain entangled with one another. Hence, neither qubit A, B, nor C has *any* definite state at all; their states are mutually interdependent but undetermined until the bit-value of one of them is measured. But determining the bit-value of any one of these qubits fixes the bit-values of the other two as well.

Note that this situation is quite different from the case of a classical probabilistic Turing machine. In that case, if we evolve the PTM without looking at the tape, the fact that we do not look has no consequence. At each random choice-point during the evolution of the PTM *some* computational path is chosen, that is, the bits A, B, and C acquire some definite bit-value, but we just don't know which one. Hence, the bits in a PTM have a definite (but possibly unknown) bit-value at all times. By contrast in the QTM, at each random choice-point *all* computational paths are pursued and, in general, the resulting state of the QTM cannot be expressed in terms of a product of a definite state for each qubit individually. Hence, the qubits in a QTM usually don't have *any* definite value until they are measured. Entanglement therefore signals a radical departure from the classical theory of computation. This makes the calculation of the net probability of a particular computational outcome different for a PTM than for a QTM.

In a quantum Turing machine (QTM) the quantum "tape" exists in a superposition of all bit string configurations, weighted with different probability amplitudes. The head, shown at the bottom of the figure, can exist in a finite number of internal states and moves back and forth along the tape, scanning the contents of each cell, writing a new symbol, entering a new internal state, and moving left or right one cell. The action of the QTM is defined by control rules that specify the probability amplitude with which each of these allowed state-to-state transitions occurs.

The tape of the QTM is infinitely long. For clarity, we zoomed in on a small snippet of it, showing the possible configurations of just three of the cells. The tape starts off in the state $a_0|\ldots 000\ldots\rangle$ $+ a_1|\ldots 001\ldots\rangle + \cdots + a_7|\ldots 111\ldots\rangle$. The "head" is initially scanning the middle cell. *If* the head were to actually read the tape, it would see a 0 with probability $|a_0 + a_1 + a_4 + a_5|^2$ and a 1 with probability $|a_2 + a_3 + a_6 + a_7|^2$. Thus, interference effects between different tape configurations can enhance or diminish the net probability of finding tape in a certain bit string configuration. However, in a QTM the head does not actually perform such a measurement. Instead, the QTM evolves coherently, without any measurements being made, into a new state $b_0|\ldots 000\ldots\rangle$ $+ b_1|\ldots 001\ldots\rangle + \cdots + b_7|\ldots 111\ldots\rangle$. The amplitudes with which each tape configuration appears in the new state reflect the possible changes that the tape could have experienced in a single step. The head could have read a 0 (or a 1) written a 0 (or a 1) and moved left (or right). Because the head did not actually perform the measurement of the tape all these computational paths must be pursued simultaneously (but with perhaps different amplitudes).

Each bit string configuration can evolve into a new configuration that differs from the original in at most the bit value in the cell that was last scanned. This restricts the pattern of allowed transitions between bit string configurations to those illustrated by the links shown in Figure 3.7. As the head can move left or right after the first step, its location becomes a superposition too. Thus, we draw the head in the second step as scanning two cells at once, and draw it in a lighter shade to suggest that its exact location is now diffusing leftward and rightward. As the head is now in several places at once, it can scan several cells, allowing more tape-to-tape transitions between the second and the third steps than between the first and the second ones. The process continues, rapidly entangling the various QTM tape configurations.

In the PTM, if a particular answer can be reached independently in more than one way, the net probability of that answer is given by the sum of all *probabilities* that lead to that answer. However, in the QTM, if a given answer can be reached in more than one way, the net probability of obtaining that answer is given by summing the *amplitudes* of all trajectories that lead to that answer and *then* squaring their absolute values to obtain the corresponding probabilities. If the quantum state of the QTM in Figure 3.3 is the superposition $c_0|00000\rangle + c_1|00001\rangle + c_2|00010\rangle + \cdots + c_{31}|11111\rangle$, then the coefficients $c_0, c_1, \ldots, c_{31}$ are the amplitudes, and the probability of finding the tape of the QTM in the bit configuration $|00010\rangle$, say, when you read each of the bits is equal to $|c_2|^2$.

Whereas classical probabilities are real numbers between zero and one, "amplitudes" are complex numbers. When we add two probabilities, we *always* get a greater or equal probability. However, when we add two complex amplitudes together, they *do not always* result in a number that has a bigger absolute value. Some pairs of amplitudes tend to cancel each other out, resulting in a net reduction in the probability of seeing a particular outcome. Other pairs of amplitudes tend to reinforce one another and thereby enhance the probability of a particular outcome. This is the phenomenon of quantum interference.

Quantum interference is a very important mechanism in quantum computing. Typically, when designing a quantum computer to solve a hard computational problem, we have to devise a method (in the form of a quantum algorithm) to cause a superposition of all the valid inputs to the problem to evolve into a superposition of all the valid solutions to that problem. If we can do so, then when we read the final state of our memory register, we are guaranteed to obtain one of the valid solutions. Understanding how to achieve this desired evolution invariably entails arranging for the computational pathways that lead to non-solutions to interfere destructively with one another, and hence cancel out, and also arranging for the computational pathways that lead to solutions to interfere constructively, and hence reinforce one another.

Armed with this model of an abstract quantum Turing machine, several researchers have been able to prove theorems about the capabilities of quantum computers. This effort has focused primarily on universality (whether one machine can simulate all others efficiently), computability (what problems the machines can do), provability (that a conjecture can be confirmed or refuted given a set of axioms and a chain of logical inferences), and complexity (how the memory and time resources scale with problem size). Let's take a look at each of these concepts and compare the perspectives given to us by classical computing and quantum computing.

Universality

The Turing machine model had a catalytic effect on computer science. In the 1930s, computer science was a fledgling field. People dabbled with building computers, but very few machines actually existed. Those that did had been tailor-made for specific applications. However, the concept of a Turing machine raised new possibilities. Turing realized that one could encode the transformation rules of any particular Turing machine, say T, as some pattern of 0s and 1s on the tape that is fed into some special Turing machine called U. U had the effect of reading in the pattern specifying the transformation rules for T and thereafter treated any further input bits exactly as T would have done. Thus U was a universal mimic of T and hence was called the Universal Turing Machine. The possibility of one machine simulating another gave a theoretical justification for pursuing the idea of a *programmable* computer.

The key issue in investigating the universality of a model computer is whether that computer can simulate any other computer and, if so, how efficiently? For example, could you use a probabilistic Turing machine to simulate a quantum Turing machine? It seems that you could by keeping track, explicitly, of all the amplitudes that define the state of the quantum computer. But how costly is it to do this? Assume that the quantum computer is to contain n qubits and that each qubit is initially in a superposition state $c_0|0\rangle + c_1|1\rangle$. Each such superposition is described by two complex numbers, c_0 and c_1, so a total of $2n$ complex numbers is needed to describe the initial state of all n qubits.

Now what happens if we want to simulate a joint operation on all n qubits? Well, the cost of the simulation skyrockets. When a joint operation is performed on all n qubits—that is, when we cause them to evolve under the action of some quantum algorithm, they are likely to become entangled with one another. Whereas the initial state could be factored into a product of states for the individual qubits, an entangled state cannot be factored in this manner. In fact, just to *write down* an arbitrary entangled state of n qubits requires 2^n complex numbers. Because a classical computer must keep track of all these complex numbers explicitly, the cost of a classical simulation of a quantum system requires a huge amount of memory and computer time.

But what about a quantum computer? Could a quantum computer simulate any quantum system efficiently? There is a good chance that it could, because the quantum computer has access to exactly the same physical phenomena as the system it is simulating.

Although in 1982 Richard Feynman gave a few examples of one quantum system simulating another, he did not prove conclusively that a universal quantum simulator was possible. But indeed it is. The question was answered in the affirmative by Seth Lloyd in 1996 (Lloyd, 1996).

This result poses something of a problem for traditional (classical) computer science. The ability to prove that all the competing models of classical computation are equivalent led Church to propose the following principle, which has subsequently become known as the Church–Turing thesis (Shapiro, 1990): "Any process that is effective or algorithmic in nature defines a mathematical function belonging to a specific well-defined class, known variously as the recursive, the λ-definable, or the Turing computable functions." Or, in Turing's words, "Every function which would naturally be regarded as computable can be computed by the universal Turing machine."

So what about simulations of quantum systems? Are they effectively computable or not? The apparent discrepancy between Feynman's observation that classical computers cannot simulate all quantum systems efficiently and the Church–Turing thesis led David Deutsch in 1985 to propose reformulating the Church–Turing thesis in physical terms. Thus Deutsch prefers "Every finitely realizable physical system can be perfectly simulated by a universal model computing machine operating by finite means." This principle can be made compatible with Feynman's observation on the efficiency of simulating quantum systems only by basing the universal model computing machine on quantum mechanics itself. This insight was the inspiration that allowed David Deutsch to prove that it was possible to devise a Universal Quantum Turing Machine, effectively a quantum Turing machine that could simulate any other quantum Turing machine efficiently. The efficiency of Deutsch's Universal Quantum Turing Machine has since been improved on by several other scientists.

Computability

Computability theory addresses the question of which problems can be solved (or which questions can be decided) in a finite amount of time on a computer. If, with respect to a particular model of computer, there is no algorithm that can guarantee to find an answer to a given problem in a finite amount of time, that answer is said to be *uncomputable* with respect to that computer. One of the great breakthroughs in classical computer science was the recognition that all of the candidate models for computers, Turing machines, recursive (self-referential) functions, and λ-definable functions were equivalent in terms of what they could and could not compute.

It was, recall, a particular question regarding computability that was the impetus behind the Turing machine idea. Hilbert's *Entscheidungsproblem* had asked whether there was a mechanical procedure for deciding the truth or falsity of any mathematical conjecture, and the Turing machine model was invented to prove that there was no such procedure.

To construct this proof, Turing used a technique called *reductio ad absurdum,* in which one begins by assuming the truth of the opposite of what one wants to prove and then derives a logical contradiction. The fact that one's assumption, coupled with purely logical reasoning, leads to a contradiction proves that the assumption must be faulty. In this case, the assumption is that there *is* a procedure for deciding the truth or falsity of any mathematical proposition, so showing that this leads to a contradiction would enable one to infer that there is in fact no such procedure.

Turing reasoned that if there *were* such a procedure, and if it were truly mechanical, then it could be executed by one of his Turing machines with an appropriate table of instructions. But a "table of instructions" can always be converted into some finite sequence of 1s and 0s. Consequently, such tables can be placed in an order, which means that the things these tables represent (the Turing machines) could also be placed in an order.

Similarly, the statement of any mathematical proposition could also be converted into a finite sequence of 1s and 0s, so they too could be placed in an order. Thus, Turing conceived of building a table whose vertical axis enumerated every possible Turing machine and whose horizontal axis enumerated every possible input to a Turing machine.

But how would a machine convey its decision on the veracity of a particular input—that is, a particular mathematical proposition? One could simply have the machine print out the result and halt. Hence the *Entscheidungsproblem* could be couched as the problem of deciding whether the *i*th Turing machine acting on the *j*th input would ever halt. Thus Hilbert's *Entscheidungsproblem* had been refashioned into Turing's Halting Problem.

Turing wanted to prove that there was no procedure by which the truth or falsity of all mathematical propositions could be decided; thus his proof begins by assuming the opposite: that there is such a procedure. Under this assumption, Turing constructed a table whose (i, j)th entry was the output of the ith Turing machine on the jth input if and only if the machine halted on that input or else some special symbol, such as $\otimes$, signifying that the corresponding Turing machine did not halt on that input. Such a table would resemble that shown in Table 3.1.

Next Turing replaced each symbol $\otimes$ with a bit, 0. The result is shown in Table 3.2.

Table 3.1
Turing's table. The ith row is the sequence of outputs of the ith Turing machine acting on inputs 0, 1, 2, 3, ...

$j \rightarrow$	0	1	2	3	4	5	6	...
i
$\downarrow$
0	$\otimes$	$\otimes$	$\otimes$	$\otimes$	$\otimes$	$\otimes$	$\otimes$	...
1	0	0	0	0	0	0	0	...
2	1	2	1	$\otimes$	3	0	$\otimes$	...
3	2	0	0	1	5	7	$\otimes$	...
4	3	$\otimes$	1	8	1	6	9	...
5	7	1	$\otimes$	$\otimes$	5	0	0	...
6	$\otimes$	2	4	1	7	3	4	...
$\vdots$	$\vdots$	$\vdots$	$\vdots$	$\vdots$	$\vdots$	$\vdots$	$\vdots$	$\ddots$

Table 3.2
Turing's table with $\otimes$ replaced by 0.

$j \rightarrow$	0	1	2	3	4	5	6	...
i
$\downarrow$
0	0	0	0	0	0	0	0	...
1	0	0	0	0	0	0	0	...
2	1	2	1	0	3	0	0	...
3	2	0	0	1	5	7	0	...
4	3	0	1	8	1	6	9	...
5	7	1	0	0	5	0	0	...
6	0	2	4	1	7	3	4	...
$\vdots$	$\vdots$	$\vdots$	$\vdots$	$\vdots$	$\vdots$	$\vdots$	$\vdots$	$\ddots$

Now because the rows enumerate all possible Turing machines and the columns enumerate all possible inputs (or, equivalently, mathematical propositions), all possible sequences of outputs—that is, all computable sequences—*must be contained somewhere in this table.* However, because any particular output is merely some number, it is possible to change each one in some systematic way—for example, by adding one to each element on a diagonal slash through the table (see Table 3.3). The sequence of outputs along the diagonal differs in the ith position from the sequence generated by the ith Turing machine acting on the ith input. Thus, this sequence cannot appear in *any* of the rows in the table. However, by the rules of its construction, the infinite table is supposed to contain *all* computable sequences, and yet here is a sequence that we can clearly compute and yet that cannot appear in any one row. Turing established a contradiction, and the assumption underpinning the argument must be wrong. That assumption was, "There exists a procedure that can decide whether a given Turing machine acting on a given input will halt." Because Turing showed that the Halting Problem is equivalent to the *Entscheidungsproblem,* demonstrating the impossibility of determining whether a given Turing machine will halt before running it shows that the *Entscheidungsproblem* must be answered in the negative, too. In other words, there is no procedure for deciding the truth or falsity of all mathematical conjectures.

This came as a bit of a shock to Hilbert and most other mathematicians, but worse was yet to come! In 1936, Kurt Gödel proved

Table 3.3

Turing's table with a 1 added to each element on the diagonal slash.

$j \rightarrow$	0	1	2	3	4	5	6	...
i								
$\downarrow$								
0	[1]	0	0	0	0	0	0	...
1	0	[1]	0	0	0	0	0	...
2	1	2	[2]	0	3	0	0	...
3	2	0	0	[2]	5	7	0	...
4	3	0	1	8	[2]	6	9	...
5	7	1	0	0	5	[1]	0	...
6	0	2	4	1	7	3	[5]	...
⋮	⋮	⋮	⋮	⋮	⋮	⋮	⋮	⋱

two important theorems that illustrated the limitations of formal systems. A formal system L is called *consistent* if we can never prove both a proposition P and its negation NOT(P) within the system. Gödel showed that "Any sufficiently strong formal system of arithmetic is incomplete if it is consistent." In other words, there are sentences P and NOT(P) such that neither P nor NOT(P) is provable using the rules of the formal system L. Because P and NOT(P) express contradictory sentences, one of them must be true, so there must be true statements of the formal system L that can never be proved. Hence Gödel showed that truth and theoremhood (or provability) are distinct concepts.

In a second theorem, Gödel showed that the simple consistency of L cannot be proved in L. Thus a formal system might be harboring deep-seated contradictions.

The results of Turing and Gödel are quite surprising. Do similar problems arise in the quantum theory of computation?

Quantum Computability

In the 1980s, some scientists began to think about the possible connections between physics and computability. The connections turn out to be rather deep (Lloyd, 1993b). For one thing, we must distinguish between nature, which does what it does, and physics, which provides models of nature expressed in mathematical form. The fact that physics is a mathematical science means that it is ultimately a formal system.

Asher Peres and Wojciech Zurek have articulated three reasonable desiderata of a physical theory (Peres, 1982): determinism, verifiability, and universality (that is, the theory can describe anything). They conclude that

> *Although quantum theory is universal, it is not closed. Anything can be described by it, but something must remain unanalyzed. This may not be a flaw of quantum theory: It is likely to emerge as a logical necessity in any theory which is self-referential, as it attempts to describe its own means of verification. In this sense it is analogous to Gödel's undecidability theorem of formal number theory: the consistency of the system of axioms cannot be verified because there are mathematical statements which can neither be proved nor disproved by the use of the formal rules of the theory, although their truth may be verified by metamathematical reasoning.*

In a later paper, Peres points out an amusing paradox (Peres, 1985). He shows that it is possible to set up three quantum observables such that two of the observables have to obey the Heisenberg Uncertainty Principle. This principle says that certain pairs of observables, such as the position and momentum of a particle, cannot be measured simultaneously with infinite precision. Measuring one such observable *necessarily* disturbs the complementary observable, so you can never measure *both* observables together. Nevertheless, Peres arranges things so that he can use the rules of quantum mechanics to predict, *with certainty,* the value of *both* of these variables individually. Thus, we arrive at an example system that we can say things about but that we can never determine experimentally — a physical analog of Gödel's undecidability theorem.

These results are all consequences of treating physics as a formal system. Is it possible to make more pointed statements about computability and quantum computers?

The first work in this area appeared in David Deutsch's original paper on quantum Turing machines (Deutsch, 1985). Deutsch argued that quantum computers could compute certain outputs, such as true random numbers, that are not computable by any deterministic Turing machine. Classical deterministic Turing machines can compute only functions — that is, mathematical procedures that return a single reproducible answer. However, there are certain computational tasks that cannot be performed by evaluating any function. For example, there is no *function* that generates a true random number. Consequently, a Turing machine can only feign the generation of random numbers.

In the same paper, Deutsch introduced the idea of quantum parallelism, which we discussed in Chapter 2. However, the power of quantum parallelism was not understood until 1991, when Richard Jozsa gave a mathematical characterization of the class of functions (joint properties) that were *computable* by quantum parallelism (Jozsa, 1991). Jozsa discovered that if a function f takes integer arguments in the range 1 to m and returns a bit value (0 or 1) output, and if the joint property function that defines some collective attribute of the outputs of f takes m bits as input and returns a single-bit output, then only a fraction $(2^{2^m} - 2^{m+1})/2^{2^m}$ of all possible joint-property functions are computable by the method of quantum parallelism. Thus, quantum parallelism alone is not going to be sufficient to solve all the joint-property questions we might wish to ask.

Of course, we could always make a QTM simulate a classical TM and compute a particular joint property in that way. Although this is feasible, it is not desirable, because the resulting computation would be no more efficient on the quantum computer than on the classical machine. However, the ability of a QTM to simulate a TM means that the class of functions computable on QTMs exactly matches the class of functions computable on classical TMs.

Proving versus Providing Proof

Many decades have passed since Turing first dreamed of his machine, and today there are a number of programs around that actually perform like artificial mathematicians in exactly the sense Turing anticipated. Current interest in these programs stems from a wish to build machines that can perform not only *mathematical* reasoning but also more general kinds of logical inference such as medical diagnosis, dialog management, and even legal reasoning. Typically, these programs consist of three distinct components: a reservoir of knowledge about some topic (in the form of axioms and rules of inference), an inference engine (which provides instructions on how to pick which rule to apply next), and a specific conjecture to be proved.

In one of the earliest examples, SHRDLU, a one-armed robot, was given a command in English that was converted into its logical equivalent and then used to create a program to orchestrate the motion of the robot arm (Winograd, 1972). Thus, simply by following rules for manipulating symbols, the robot gave the appearance of understanding a command in plain English.

In a more contemporary example, the British Nationality Act was encoded in first-order logic, and a theorem prover was used to uncover logical inconsistencies in the legislation. Similarly, the form of certain legal arguments can be represented in logic that can then be used to find precedents by revealing analogies between the current case and past examples.

Although most people would think themselves far removed from the issue of "theorem proving," they could be in for a surprise if the tax authorities decided to play these games with the tax laws!

Today's artificial mathematicians are far less ingenious than their human counterparts. On the other hand, they are infinitely more patient and diligent. These qualities can sometimes allow artificial mathematicians to churn through proofs upon which no human would have dared embark. Take, for example, the case of map coloring. Cartographers had long conjectured that with just four different colors, they could color any planar map so that no two adjacent regions were the same color. However, this conjecture resisted all attempts to construct a proof for many years. In 1976, the problem was finally solved with the help of an artificial mathematician. The "proof" was unusual in that it ran some 200 pages! For a human even to check it, let alone generate it, would be a mammoth undertaking. Table 3.4 lists some milestones in mathematical proof by humans and machines.

Despite differences in the "naturalness" of the proofs they find, artificial mathematicians are nevertheless similar to real mathematicians in one important respect: Their output is an explicit sequence of reasoning steps (a proof) that, if followed meticulously, would convince a skeptic that the information in the premises, combined with the rules of logical inference, would be sufficient for us to deduce the conclusion. Once such a chain is found, the theorem has been proved. The important point is that the proof chain is a tangible object that can be inspected at leisure.

Surprisingly, this is not necessarily the case with a QTM. In principle, a QTM could be used to create some proof that relied on quantum–mechanical interference among all the computations go-

Table 3.4
Some impressive mathematical proofs.

Mathematician	Proof Feat	Notable Features
Daniel Gorenstein (1980)	Classification of finite simple groups	Created by human. 15,000 pages long.
Kenneth Appel and Wolfgang Haken (1976)	Proved the four-color theorem	Created by computer. Reduced all planar maps to combinations of 2000 special cases and then exhaustively checked each case.
Andrew Wiles (1993)	Proved Fermat's last theorem	Created by human. 200 pages long. Only 0.1 percent of all mathematicians are competent to judge its veracity.
Laszlo Babai and colleagues	Invented probabilistic proof checking	Able to verify that a complex proof is "probably correct" by replicating any error in the proof in many places in the proof, thereby amplifying the chances of the error being detected.

ing on in superposition. Upon interrogating the QTM for an answer, we might be told, "Yes, your conjecture is true," but there would be no way to exhibit all the computations that the machine had performed to arrive at the conclusion. Thus, for a QTM, the ability to prove something and the ability to provide the proof trace are quite distinct concepts.

Worse still, if we tried to peek inside the QTM as it was working, to glean some information about the state of the proof at that time, we would inevitably disrupt the future course of the proof.

Complexity

Whereas "computability" concerns which computational problems computers can and cannot do, "complexity" concerns how efficiently problems can be solved. Efficiency is an important consideration for real-world computing. The fact that a computer can, in principle, solve a particular kind of problem does not guarantee that it can solve it in practice. If the running time of the computer is too long or the memory requirements too great, then an apparently feasible computation can still lie beyond the reach of any practicable computer.

Computer scientists have developed a rich classification scheme for describing the complexity of various algorithms running on different kinds of computers. The most common measures of efficiency employ the rate of growth of the time or memory needed to solve a problem as the size of the problem increases. Of course, *size* is an ambiguous term. Loosely speaking, the size of a problem is taken to be the number of bits needed to state the problem to the computer. For example, if an algorithm is being used to factor a large integer N, then the size of the integer being factored would be roughly $\log_2 N$.

The reason why complexity classifications are based on the rates of growth of running times and memory requirements, rather than on absolute running times and memory requirements, is to factor out the variations in performance experienced by different makes of computers with different amounts of RAM, swap space, and processor speeds. Using a growth-rate–based classification makes the complexity of a particular algorithm an intrinsic measure of the difficulty of the problem the algorithm addresses.

Although complexity measures are independent of the precise make and configuration of computer, they are related to a particular mathematical *model* of the computer, such as a deterministic Turing machine or a probabilistic Turing machine. It is now known, for example, that many problems that are intractable with respect to a deterministic Turing machine can be solved efficiently, with high probability, on a probabilistic Turing machine.

There are many criteria by which we could assess how efficiently a given algorithm solves a given type of problem. For the better part of the century, computer scientists focused on worst-case complexity analyses. These have the advantage that if an efficient algorithm can be found for solving some problem *in the worst case*, then it is sure to be an efficient algorithm for *any* instance of such a problem.

This estimation can be somewhat misleading, however. Recently some computer scientists have developed average-case complexity analyses. Moreover, it is possible to understand the finer-grain structure of complexity classes and locate regions of especially hard and especially easy problems within a supposedly "hard" class (Williams, 1994). Nevertheless, one of the key questions is whether some algorithm runs in polynomial time or exponential time. Thus, we'd better explain what these terms mean.

Polynomial versus Exponential Growth

Computer scientists have developed a rigorous way of quantifying the difficulty of a given type of problem. This classification is based on the mathematical form of the function that describes how the computational cost incurred in solving the problem *scales up* as larger problems are considered. The most important quantitative distinction is between polynomially growing costs (which are deemed tractable) and exponentially growing costs (which are deemed intractable). The notion of "cost" can include the time taken to complete the computation, the amount of memory space consumed, or a combination of both.

Exponential growth will always exceed polynomial growth eventually, regardless of the order of the polynomial. For example, Figure 3.4 compares the growth of the exponential function $\exp(L)$ with the growth of the polynomials L^2, L^3, and L^4. As you can see, whatever the degree of the polynomial in L, the exponential eventually becomes larger.

Multiplication and factoring are a good pair of problems to illustrate the radical difference between polynomial and exponential growth. It is relatively easy to multiply two large numbers together

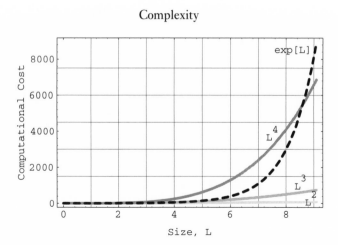

Figure 3.4 Comparison of exponential and polynomial growth rates. An exponential always beats a polynomial eventually.

to obtain their product, but it is extremely difficult to do the opposite—to find the factors of a composite number:

$$10433 \times 16453 = ? \quad \text{(easy)}$$

$$? \times ? = 200949083 \quad \text{(hard)}$$

If, in binary notation, the numbers being multiplied have L bits, then multiplication can be done in a time proportional to L^2, a polynomial in L.

For factoring, the best-known classical algorithms are the Multiple Polynomial Quadratic Sieve (Silverman, 1987), for numbers involving roughly 100–150 decimal digits, and the Number Field Sieve (Lenstra, 1990), for numbers involving more than roughly 110 decimal digits. The running time of these algorithms grows subexponentially (but superpolynomially) in L, the number of bits needed to specify the number to be factored, N. The best factoring algorithms require a time on the order of $\exp(L^{1/3} \log(L)^{2/3})$. Exponential, subexponential, and polynomial growth rates are compared in Figure 3.5.

Richard Crandall has charted the progress in feats of factoring over the past three decades (see Table 3.5; Crandall, 1996). In the early 1960s, computers and algorithms were good enough only to factor numbers with 20 decimal digits; by 1994 that limit had risen to 129-digit numbers, but only after a Herculean effort.

The presumed difficulty of factoring large integers is the basis for the security of the so-called public key cryptosystems that are in

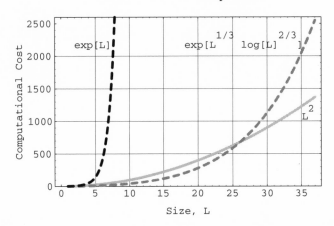

Figure 3.5 The best factoring algorithms grow subexponentially (but super-polynomially) in L, the number of bits needed to specify the number being factored.

widespread use today. When one of these systems was invented, the authors offered a prize to anyone who could factor the following 129-digit number (called RSA 129).

$$RSA129 = 11438162575788886766923577997614661120102182$$
$$9672124236256256184293570693524573389783059$$
$$7123563958705058989075147599290026879543541$$

In 1994, a team of computer scientists using a network of workstations succeeding in factoring RSA129. The resulting factors are

$$3490529510847650949147849619903898$$
$$13341776463849338784399082057\ 7$$

Table 3.5
Progress in factoring large composite integers. One MIP-year is the computational effort of a machine running at 1 million instructions per second for 1 year.

Year	Factorizable	Effort (MIP-years)
1964	20 decimal digits	0.000009
1974	45 decimal digits	0.001
1984	71 decimal digits	0.1
1994	129 decimal digits	5000

and

$$3276913299326670954996199881908344$$
$$6141317764296799294253979828\8533$$

Extrapolating the observed trend in factoring suggests that it would take about 2.9 billion MIP-years to factor a 200-digit number.

Classical Complexity Classes

Knowing the exact functional forms for the rates of growth of the number of computational steps for various algorithms allows computer scientists to classify computational problems on the basis of difficulty. The most useful distinctions are based on classes of problems that either can or cannot be solved in polynomial time, in the worst case. Problems that can be solved in polynomial time are usually deemed to be tractable and are lumped together into the class P. Problems that cannot be solved in polynomial time are usually deemed to be intractable and may be in one of several classes. Of course, it is possible that the order of the polynomial is large, such as 12, which makes solving a supposedly tractable problem rather difficult in practice. Fortunately, such large polynomial growth rates do not arise often, and the distinction between polynomial and exponential time is a pretty good indicator of difficulty.

The next most interesting class after P is the class NP. Many algorithms for tackling a variety of practical problems involve trying to extend partial solutions into bigger partial solutions that eventually lead to a full solution. If a partial solution is found that cannot be extended further, the algorithm has to back up and try another alternative. Unfortunately, the number of computational steps required to guarantee that a deterministic algorithm will find a solution is often exponential in the size of the problem. If so, that type of problem is effectively intractable. However, many such problems also have the feature that once a candidate solution has been found, its correctness can be tested efficiently—that is, in polynomial time. Thus, if we could magically guess the right solution, the problem *could* be solved efficiently. Such a possibility means that there is, in principle, an efficient *non-deterministic* algorithm for solving the problem. Computer scientists therefore lump all such problems into a complexity class called NP, which stands for "non-deterministic polynomial time."

One question of great interest to computer scientists is whether the class P is the same as the class NP—that is, whether $P = NP$. So far, nobody knows.

Within the class NP there is a special subclass of problems, called NP-complete, that has a special property: Any NP problem can be converted into an NP-complete problem in polynomial time. Likewise, any NP-complete problem can be converted into any other NP-complete problem in polynomial time. Thus, if anyone *could* solve one of the NP-complete problems efficiently (i.e., in polynomial time) then it follows that *all* problems in NP could be solved efficiently. Thus, the fate of one NP-complete problem is intimately bound to that of its kin. Either all NP-complete problems are tractable or none of them are! They stand or fall together. Currently, around 1000 distinct NP-complete problems are known. They crop up in many real-world problems, such as scheduling, route planning, and cargo packing.

Are NP-complete problems the hardest problems of all? Not by a long shot. There is a towering hierarchy of monstrous problems whose running times are doubly, triply, or quadruply exponential in the size of the input; that is, their running times grow as 2^{2^N} or $2^{2^{2^N}}$ and so on. One example is Presburger arithmetic, a superficially innocuous-looking logic that allows mathematicians to prove properties about formulas built out of positive integers and variables whose values can be positive integers. Presburger arithmetic is *at best* doubly exponential.

Even though NP-complete problems are not the hardest problems of all, they crop up in practical applications so often that there is a desperate need for efficient algorithms to tackle them. Consequently, it would be a great achievement if someone could show how to make a quantum computer solve an NP-complete problem more efficiently than a classical (deterministic or probabilistic) computer. This goal is being pursued actively by several researchers (Cerf, Grover & Williams, 1998). Table 3.6 lists and describes some classical complexity classes.

One approach is to recognize that NP-complete problems share a special *structure* that allows complete solutions to be assembled out of smaller, partial solutions (Williams and Hogg, 1994). For example, suppose you have to decide whether it is possible to color the nodes of an n-vertex, e-edge graph using only three colors, in such a way that no two vertices connected by an edge share the same color. This kind of "graph coloring" problem is in the family of NP-complete problems. As the size of the problem considered is increased,

Table 3.6
Some classical complexity classes.

Classical Complexity Class	Intuitive Meaning	Examples
P (or PTIME)	Polynomial-time: the running time of the algorithm is, in the worst case, a polynomial in the size of the input. All problems in P are tractable.	Multiplication
NP	Non-deterministic polynomial time: a candidate answer can be checked for correctness in polynomial time.	Factoring composite integers
NP-complete	All problems in NP can be mapped into any of the NP-complete problems in polynomial time. If just one of the problems in the NP-complete class is shown to be tractable, then problems in NP must all be tractable.	Scheduling Satisfiability Traveling salesman problem
ZPP	Can be solved, with certainty, by PTMs in average case polynomial time.	Randomized Quicksort
BPP	Can be solved in polynomial time by PTMs with probability $> 2/3$. Probability of success can be made arbitrarily close to 1 by iterating the algorithm a certain number of times.	Problems in BPP are tractable Decision version of Min-Cut

there is no known classical algorithm that can *guarantee* to decide, within a time that grows polynomially with the size of the problem, whether a given graph can be colored using three colors. An example of such a graph coloring problem is shown in Figure 3.6.

For graph coloring, a partial assignment specifies colors for just some of the vertices in the graph. Partial assignments that violate one or more constraints are impossible to extend into complete solutions. However, partial assignments that satisfy all of the constraints against which they may be tested are called *partial solutions*

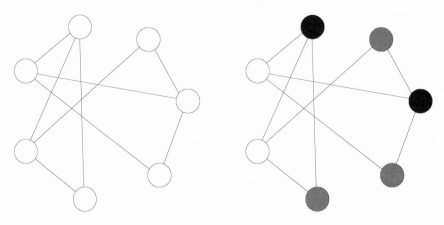

Figure 3.6 A graph with seven nodes and nine edges which can be colored using only three colors, namely, white, gray, or black.

and have the *possibility* of being extended into complete solutions. Thus, one way to construct complete solutions is by trying out different extensions of partial solutions. To make such an extension you must pick a currently uncolored node, assign it a color, and check that the new assignment is compatible with the previous color assignments and with the constraints due to the edges. Eventually, you will either reach a dead end and be forced to back up and revise some earlier color choices, or you will reach a complete solution and hence decide that the graph is 3-colorable. If you eliminate all possible color assignments for any vertex, then the decision is that the graph is not 3-colorable.

This process of extending and retracting partial solutions can be pictured as a search in a tree of partial solutions. In such a tree, the nodes at the ith level correspond to all possible partial assignments of colors to i out of the n vertices. The depth of the tree equals the number of vertices in the graph being colored, and the number of branches emanating from each node in the tree, is the number of colors that you are allowed to use.

A quantum algorithm for performing search in such a tree structure is illustrated in Figure 3.7. This algorithm works by nesting a quantum search in the middle of the tree with another in the fringe. This is a bit like checking all the partial assignments of a given size and pursuing only those that seem promising.

The net effect is to bias the search in the fringe of the tree in favor of only the extensions of the partial solutions, that is, partial as-

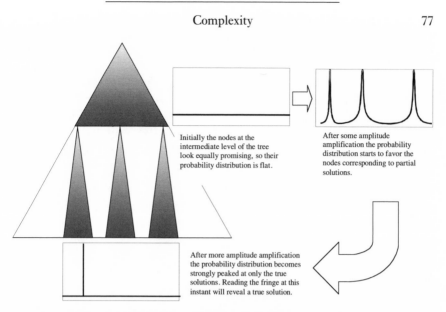

Figure 3.7 A nested quantum search embeds a quantum search at an intermediate level of the search tree with another in the fringe of the tree.

signments that appear to be viable, at the intermediate level. If there are N candidates in the fringe of the tree, then the nested quantum search can find a solution amongst them in roughly $\sqrt[3]{N}$ steps for the hardest problem instances (i.e., the graphs that have just enough edges to eliminate all but one solution).

By way of setting things up for the future, we should point out the classical complexity behavior of two important problems: primality testing and factoring. These problems arise in RSA public key cryptography, a scheme for sending and receiving secret coded messages that are hard to decode using a classical computer unless you know a special number called a key. We explain the RSA public key cryptography scheme fully in Chapter 4. Primality testing (via probabilistic algorithms) is fast. In RSA, the recipient must choose two large prime numbers. If proving that a number is prime were slow, then RSA would not be feasible. Factoring, however, is not fast on any type of classical computer; that is, so far no one has found an algorithm, deterministic or probabilistic, that can factor a number in polynomial time. In RSA cryptography, an eavesdropper must factor a publicly known number in order to obtain the private key. If factoring were easy, RSA would be insecure. We shall have more to say about factoring on a quantum computer in Chapter 4.

Quantum Complexity Classes

Just as there are classical complexity classes, there are quantum complexity classes. Because quantum Turing machines are quantum-mechanical generalizations of probabilistic Turing machines, the quantum complexity classes resemble the probabilistic complexity classes. There is a tradeoff between the certainty of the answer being correct and the certainty of the answer being available within a given time. In particular, the classical classes P, ZPP, and BPP (see Table 3.4) become the quantum classes QP, ZQP, and BQP. These mean that a problem can be solved with certainty in worst-case polynomial time, with certainty in average-case polynomial time, and with probability greater than 2/3 in worst-case polynomial time, respectively, by a quantum Turing machine.

Statements about the relative power of one type of computer compared with another can be couched in the form of subset relationships among complexity classes. Thus, QP is the class of problems that can be solved, with certainty, in polynomial time on a quantum computer, and P is the set of problems that can be solved, with certainty, in polynomial time on a classical computer. Because the class QP contains the class P (see Table 3.7), this means that there are more problems that can be solved efficiently by a quantum computer than by any classical computer. Similar relationships are now known for some of the other complexity classes, but many open questions remain.

The study of quantum complexity classes began with David Deutsch in his original paper on quantum Turing machines (QTMs). The development of the field is summarized in Table 3.8.

In Deutsch's original paper, he presented the idea of quantum parallelism. Quantum parallelism enables us to compute an exponential number of function evaluations in the time it takes to do just one function evaluation classically. Unfortunately, the laws of quantum mechanics make it impossible to extract more than one of these answers explicitly. The problem is that although we can indeed calculate all the function values for all possible inputs at once, when we read off the final answer from the tape, we obtain only one of the many outputs. Worse still, in the process, the information about all the other outputs is lost irretrievably. Thus, the net effect is that we are no better off than if we had used a classical Turing machine. As far as function evaluation goes, then, the quantum computer is no better than a classical computer.

Table 3.7
Some new, quantum complexity classes and their known relationships to classical complexity classes.

Quantum Class	Class of Computational Problems That Can . . .	Relationships to Classical Complexity Classes (if known)
QP	. . . be solved, with certainty, in worst-case polynomial time by a quantum computer.	P ⊂ QP (The quantum computer can solve more problems in worst-case polynomial time than can the classical computer.)
BQP	. . . be solved in worst-case polynomial time by a quantum computer with probability > 2/3 (thus the probability of error is bounded; hence the B in BQP).	BPP ⊆ BQP ⊆ PSPACE (That is, it is not known whether QTMs are more powerful than PTMs.) Shor proved factoring is in BQP.
ZQP	. . . solved with zero error probability in expected polynomial time by a quantum computer.	ZPP ⊂ ZQP

Deutsch realized that it was possible to calculate certain *joint properties* of all the answers without having to reveal any one answer explicitly. (We explained how this works at the end of Chapter 2.) Richard Jozsa refined Deutsch's ideas about quantum parallelism by showing that many functions—for example, SAT, the propositional satisfiability problem—cannot be computed by quantum parallelism at all (Jozsa, 1991). Nevertheless, the question of the utility of quantum parallelism for tackling computational tasks that were *not* function evaluations remained open.

In 1992, Deutsch and Jozsa proposed a problem that was not equivalent to a function evaluation and for which a quantum Turing machine (QTM) was exponentially faster than a classical deterministic Turing machine (DTM). The problem was rather contrived and consisted of finding a true statement in a list of two statements. It was possible that both statements were true, in which case either statement would have been acceptable as the answer. This potential multiplicity of solutions meant that the problem could not be refor-

text continues on page 82

Table 3.8
Historical development of quantum complexity theory.

Year	Advance in Quantum Complexity Theory
Benioff (1980)	Shows how to use quantum mechanics to implement a Turing machine (TM).
Feynman (1982)	Shows that TMs cannot simulate quantum mechanics without exponential slowdown.
Deutsch (1985)	Proposes first universal QTM and the method of quantum parallelism.
	Proves that QTMs have the same complexity class with respect to function evaluation as TMs. Remarks that certain computational tasks do not require function evaluation. Exhibits such a task that is solved faster on a QTM than on a TM.
Jozsa (1991)	Describes classes of functions that can and cannot be computed by quantum parallelism.
Deutsch and Jozsa (1992)	Exhibit a contrived problem that the QTM solves with certainty in poly-log time, but that requires linear time on a DTM. Thus, the QTM is exponentially faster than the DTM. Unfortunately, the problem is also easy for a PTM, so this is not a complete victory over classical machines.
Berthiaume and Brassard (1992)	Prove $P \subset QP$ (strict inclusion). The first definitive complexity separation between classical and quantum computers.
Bernstein and Vazirani (1993)	Describe a universal QTM that can simulate any other QTM efficiently (Deutsch's QTM could simulate other QTMs, but only with an exponential slowdown).
	Show how to sample from the Fourier spectrum of a Boolean function on n bits in polynomial time on a QTM (set up for Simon's paper and Shor's paper).
Yao (1993)	Shows that complexity theory for quantum circuits matches that of QTMs. This legitimizes the study of quantum circuits (which are simpler to design and analyze than QTMs).
Berthiaume and Brassard (1994)	Prove that randomness alone is not what gives QTMs the edge over TMs.
	Prove that there is a decision problem that is solved in polynomial time by a QTM, but requires exponential time, in the worst case, on a DTM or PTM. First time anyone showed a QTM to beat a PTM.

(continued)

Table 3.8
Continued.

Year	Advance in Quantum Complexity Theory
Berthiaume and Brassard (1994) *(continued)*	Prove that there is a decision problem that is solved in exponential time on a QTM, but requires *double* exponential times on a DTM in all but a few instances.
Simon (1994)	Lays foundation for Shor's algorithm.
Shor (1994)	Discovers a polynomial-time quantum algorithm for factoring large integers. This is the first *significant* problem for which a quantum computer is shown to outperform any type of classical computer. Factoring is related to breaking codes in widespread use today.
Grover (1996)	Discovers a quantum algorithm for finding a single item in an unsorted database in the square root of the time it would take on a classical computer. The problem takes N steps classically and $(\pi/4)\sqrt{N}$ quantum–mechanically.
Grover (1997)	Can find an item in one step, provided the query is sufficiently complicated.
Fijany and Williams (1998)	Demonstrate quantum circuit for the quantum wavelet transform.
Grover (1998)	Can use almost any unitary operator for quantum search.
Cerf, Grover, and Williams (1998)	Quantum speedup of NP-complete problems by nesting one quantum search within another. The problem takes N^x steps classically and $\sqrt{N^x}$ steps quantum-mechanically (with $x < 1$ depending on degree of constrainedness of the problem).
Brassard & Hoyer	Adapt Grover and Shor's algorithms for quantum-counting the number of solutions to a problem.
Abrams & Lloyd (1998)	Present a polynomial time quantum algorithm for finding eigenvalues and eigenvectors of certain matrices.
W. van Dam (1998)	Shows how a quantum computer can guess an N-bit string (known to a quantum oracle) in roughly $N/2 + \sqrt{N}$ calls to the oracle.
Beals et al. (1998)	Prove bounds on the efficiency of quantum computers relative to classical deterministic computers for several computational problems. In many cases if the quantum machine takes T steps the classical one takes at most $O(T^6)$ steps.

mulated as a function evaluation. The upshot was that the QTM could solve the problem in "polynomial in the logarithm of the problem size" time (poly-log time), but that the DTM required linear time. Thus the QTM was exponentially faster than the DTM. The result was only a partial success, however, because a probabilistic Turing machine (PTM) could solve the problem as efficiently as the QTM. But this did show that a quantum computer at least could beat a deterministic classical computer.

Now the race was on to find a problem of which the QTM beat a DTM *and* a PTM. Ethan Bernstein and Umesh Vazirani (Bernstein, 1993) analyzed the computational power of a QTM and found a problem for which it did beat both a DTM and a PTM. Given any Boolean function on *n*-bits, Bernstein and Vazirani showed how to sample from the Fourier spectrum of the function in polynomial time on a QTM. It was not known whether this was possible on a PTM. This was the first result that hinted that QTMs might be more powerful than PTMs.

The superiority of the QTM was finally clinched by André Berthiaume and Gilles Brassard, who constructed an "oracle" relative to which there was a decision problem that could be solved with certainty in worst-case polynomial time on the quantum computer, yet could not be solved classically in probabilistic expected polynomial time (if errors were not tolerated). Moreover, they also showed that there is a decision problem that can be solved in exponential time on the quantum computer and that requires double exponential time on all but finitely many instances on any classical deterministic computer. This result was proof that a quantum computer could beat both a deterministic and a probabilistic classical computer, but it was still not headline news because the problems for which the quantum computer was better were all rather contrived.

The situation changed when, in 1994, Peter Shor, building on work by Dan Simon, devised polynomial-time algorithms for factoring integers and for computing "discrete logarithms." The latter two problems are believed to be intractable for any classical computer, deterministic or probabilistic. But more important, the factoring problem is intimately connected with the ability to break the RSA cryptosystem that is in widespread use today. Thus, if a quantum computer could break RSA, then a great deal of sensitive information would suddenly become vulnerable, at least in principle. Whether it is vulnerable in practice depends, of course, on the feasibility of designs for actual quantum computers.

Of course, computer scientists would like to develop a repertoire of quantum algorithms that can, in principle, solve significant computational problems faster than any classical algorithm. Unfortunately, the discovery of Shor's algorithm for factoring large composite integers was not followed by a wave of new quantum algorithms for lots of other problems. To date, only about a dozen quantum algorithms are known. These are the algorithms of Deutsch/Jozsa, true-statement problem (Deutsch, 1992); Simon (Simon, 1994); Shor, factoring (Shor, 1994); Kitaev, factoring (Kitaev, 1995); Grover, database search (Grover, 1996) and estimating the median (Grover, 1996a); and Durr/Hoyer, estimating the mean (Durr, 1996). Many of these quantum algorithms rely on a quantum version of the Fourier transform.

Searching a Quantum Phone Book

One of the most versatile quantum algorithms to have been discovered in recent years is Lov Grover's quantum algorithm for "unstructured quantum search" (Grover, 1996). A more intuitive way to think of this problem is in terms of looking up someone's name in a telephone directory, given that you only know his or her telephone number. So we will use the telephone directory analogy as an example.

As you probably know, it is not that hard to find someone's number in a telephone directory if you know the person's name. This is because the telephone directory is indexed alphabetically by name. One strategy for looking up a number, given that you know the name, is to open the directory at a random page and ask yourself whether the name occurs before or after the page you selected. If the name occurs before the opened page then open another page at random in the earlier part of the directory. If the name occurs after the opened page, open a page at random in the latter part of the directory. Repeating this process just a handful of times, allows you to hone in on the sought after number in just a few tries. We can associate a computational complexity with such a search process. It turns out that the complexity scales roughly logarithmically with the number of entries in the telephone directory. So a telephone directory containing N names can be searched in roughly $\log_2 N$ steps.

The inverse of this search problem—that of finding the name of someone who has a prescribed telephone number—is substantially more difficult. This is because the telephone directory is essentially

unsorted with respect to telephone numbers (as it is sorted alphabetically with respect to names, and it can't be sorted by both criteria
simultaneously). The only way to search for the name, given the
number, amounts to trial and error, examining all the numbers in a
random order. Thus, if the telephone directory contains N numbers
you can expect to find the name corresponding to a prescribed number in roughly $N/2$ steps on average and in N steps in the worst
case. This means that the problem of looking up a name given a
number is exponentially more difficult than that of looking up a
number given a name.

But what if your telephone directory was replaced by a quantum
telephone directory "oracle" that could answer any question of the
form "does the ith indexed entry correspond to the given number?"
with a simple "yes" or "no"? Could you search such a quantum
telephone directory for someone's name, given his or her number,
more efficiently than you could classically?

The answer turns out to be a surprising "yes." It is surprising
because on the face of it there does not seem to be any hope of obtaining much guidance from the failures you encounter during the
search process. But the miracle of quantum computing that makes
an improvement possible is the fact that you are allowed to pose a
superposition of questions to the oracle, obtaining a superposition
of replies. This is yet another example of the phenomenon of quantum parallelism that was discussed in Chapter 2.

The basic scheme of Grover's algorithm is illustrated in Figure
3.8. We start with a uniform superposition of the possible indices of
the entries in the telephone directory. Somewhere in the midst of this
superposition is the target index that we are seeking. In Figure 3.8
this is illustrated by the little stick man squeaking "yes" when all his
colleagues are squeaking "no." But everyone is shouting equally
loudly.

To perform quantum search, we are going to use the quantum
telephone directory oracle to create a special amplitude amplification operator called $\hat{Q}$ in the figure. Applying $\hat{Q}$ a certain number of
times has the effect of boosting the amplitude of the "yes" stick man
at the expense of the amplitudes of the others. So the "yes" stick
man ends up shouting louder than his colleagues, making him much
easier to pick out from the crowd.

For simplicity, let's suppose the quantum telephone directory
happens to contain $N = 2^n$ entries. (Grover's algorithm can be gen-
eralized to the case when this is not true, but it is simpler to think

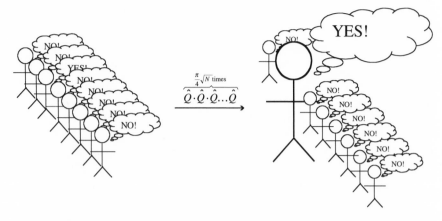

Figure 3.8 Illustration of Grover's unstructured quantum search algorithm.

about the algorithm when N is a power of 2.) If you start with a quantum state representing a superposition of all possible indices of the entries in your telephone directory, you can ask the quantum telephone directory oracle to pronounce on whether each of them is the sought-after index simultaneously. By way of reply, the oracle hands you the state you submitted in your query modified so that the phase of the sought-after index is flipped, but those of the other indices are not. Thus, the oracle "tags" the target index by way of a controlled phase inversion.

For example, suppose you had a quantum phone book that had four entries (i.e., $N = 4$). Imagine that the target index is $|t\rangle = |10\rangle$. In other words, the name corresponding to your given number is actually the third entry in this telephone directory. The initial query you would submit to the quantum telephone book oracle would be the quantum state $(1/\sqrt{4})(|00\rangle + |01\rangle + |10\rangle + |11\rangle)$. This query is asking the oracle implicitly whether each of the indices (eigenstates) $|00\rangle$, $|01\rangle$, $|10\rangle$, and $|11\rangle$ is the index of the target state. The oracle replies with the state $(1/\sqrt{4})(|00\rangle + |01\rangle - |10\rangle + |11\rangle)$. This is the same as the query state except that the sign of the third index, that is, the sign of the target, has been flipped.

Given such an oracle, a starting state $|s\rangle$, and an arbitrary unitary operator $\hat{U}$, Lov Grover showed how to create an amplitude amplification operator $\hat{Q} = -\hat{U} \cdot \hat{I}_s \cdot \hat{U}^{-1} \cdot \hat{I}_t$ where $\hat{I}_s = \hat{1} - 2|s\rangle\langle s|$ and $\hat{I}_t = \hat{1} - 2|t\rangle\langle t|$, which has the effect of slightly boosting the amplitude of the target index $|t\rangle$ each time $\hat{Q}$ is applied. Although it looks as though you need to know $|t\rangle$ in order to construct the amplitude amplification operator to find $|t\rangle$, in reality you do not because the *effect* of $\hat{I}_t$ can be achieved using the quantum oracle. So it is not necessary to know the answer to the problem in order to solve the problem, which is a requirement for a legitimate computation!

A good choice for $\hat{U}$ is the n-qubit Walsh–Hadamard operator. The 1-qubit Walsh–Hadamard operator performs the rotations $|0\rangle \mapsto (1/\sqrt{2})(|0\rangle + |1\rangle)$ and $|1\rangle \mapsto (1/\sqrt{2})(|0\rangle - |1\rangle)$. The n-qubit version consists of applying the 1-qubit Walsh–Hadamard operator to n qubits simultaneously. Thus, starting with the state $|s\rangle = |00\ldots0\rangle$ in which all n qubits are $|0\rangle$, the n-qubit Walsh–Hadamard operator creates a uniformly weighted superposition of all indices in the range 0 to 2^n, that is, $(1/\sqrt{N}) \times (|00\ldots0\rangle + |00\ldots1\rangle + \cdots + |11\ldots1\rangle)$. Because each index appears with exactly the same amplitude in this superposition, you can be certain that the overlap between the target state and the superposition state must be $1/\sqrt{N}$ regardless of what the identity of the target index. This is why the n-qubit Walsh–Hadamard gate is a good choice for $\hat{U}$.

If you apply the amplitude amplification operator, $\hat{Q}$, k times to the starting state $\hat{U}|s\rangle$, the amplitude of the target index grows initially by a factor of about $(1 + 2k)$. If you apply $\hat{Q}$ too many times, the amplitude begins to decreases it again. Thereafter it will vary periodically with increasing k. In fact, the overlap between the target state the amplitude amplified state roughly takes the form:

$$\langle t|\hat{Q}^k\hat{U}|s\rangle \approx \frac{1}{\sqrt{N}}\cos\left(\frac{2k}{\sqrt{N}}\right) + \sin\left(\frac{2k}{\sqrt{N}}\right).$$

Hence, the probability of obtaining the target index upon measuring the amplitude-amplified superposition is given by squaring the absolute value of this expression. The resulting oscillating probability is illustrated in Figure 3.9. The horizontal axis shows increasing values of k, that is, increasing amounts of amplitude amplification from left to right. The vertical axis shows increasing numbers of solutions to the problem, that is, there might be more than one name in the telephone directory that matches the given telephone number. The shading shows the probability of obtaining the target index. A bright white point means probability 1 (certainty) and a black point means probability 0 (impossibility). If you choose a particular value for the vertical axis and sweep out a line horizontally to the right, you will see that the probability of success rises and falls periodically with increasing amounts of amplitude amplification.

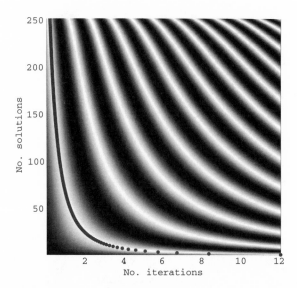

Figure 3.9 Variation in the overlap between the target state and the amplitude-amplified state as k is varied (horizontal axis) and as the number of names that all match the same phone number is varied (vertical axis).

Provided that k is much less than $\sqrt{N}$, that is, provided that you don't amplitude-amplify too far, the overlap between the target state and the amplitude-amplified state is close to $(1 + 2k)/\sqrt{N}$. Hence the probability of obtaining the target index if you were to measure the amplitude amplified superposition grows by a factor of about $(1 + 2k)^2$, that is, *quadratically* with the number of times you iterate $\hat{Q}$. Thus, to be sure to find the target, all you need to do is to amplitude-amplify the target index until its amplitude is close to 1. This can be accomplished by picking k, the number of amplitude amplification steps, to be about in $(\pi/4)\sqrt{N}$ iterations of $\hat{Q}$.

Each application of the amplitude-amplification operator, $\hat{Q}$, only requires *one* call to the quantum telephone oracle (equivalent to one lookup in the telephone directory), even though the oracle is pronouncing on the correctness of all the possible indices simultaneously. Thus the total number of calls to the quantum telephone directory is on the order of $(\pi/4)\sqrt{N}$. This is much more efficient than the order N lookups we needed to do the same task classically. Hence it marks a great improvement over what can be done classically in the same memory and time resources. Grover's algorithm is summarized in Table 3.9.

Table 3.9
Grover's Algorithm

Step 1:	Start with an equally weighted superposition of all $N = 2^n$ possible indices. Any one of which could be the target entry in the quantum telephone directory.
Step 2:	Pick an (almost) arbitrary unitary operator. The operator has to have some non-zero overlap between the starting state and the target. The easiest way to ensure this is to pick an operator with no zero entries in its unitary matrix.
Step 3:	Construct a special amplitude-amplification operator, $\hat{Q}$, from the quantum telephone directory oracle and the arbitrary unitary operator.
Step 4:	Iterate $\hat{Q}$ about $(\pi/4)\sqrt{N}$ times starting with the state $\hat{U}|s\rangle$ and then measure. The measurement outcome is the target index, with probability ≈ 1 (i.e., near certainty!)

Grover's algorithm and Shor's algorithm are the most interesting quantum algorithms to date. In the next chapter, we will explain how Shor's algorithm is related to code breaking. It was this application that was responsible for igniting interest in quantum computing.

Four

—◄►◄◆►—

Breaking
"Unbreakable" Codes

Nothing is secret which shall not be made manifest.
—Luke:8 –17

The first computational tasks for which quantum computers were found to outperform classical computers were artificial problems that had been specially contrived to demonstrate a quantum advantage. A good example is the "Deutsch–Jozsa" problem, which was described in Chapter 2. As you may recall, this problem involves deciding whether a mystery n-bit function is constant (i.e., gives the same output on all possible inputs) or balanced (i.e., gives an output of 0 on half of the possible inputs and an output of 1 on the other half). This problem does not arise in practical computing applications very often! This lack of applicability in part explains why the field of quantum computing languished for almost a decade. For quantum computing to become a hot research area, someone needed to find at least one important application for which a quantum computer outperformed any classical computer.

There is no shortage of computational tasks that we would like to perform faster. The problems of devising efficient airline schedules, planning the layout of microchips, and optimizing the design of cars are all examples of computational tasks that become more difficult to solve as the size of the problem is increased. "Size" for our examples, could be taken to be the number of aircraft or routes, the number of microchip components, or the number of design parameters.

But the breakthrough did not come in any of these areas. Instead, it arose in "cryptanalysis," an abstract area of mathematics that might at first seem far removed from everyday experience. In 1994, Peter Shor, of AT&T Bell Labs, discovered how to make a quantum computer factor a large integer super-efficiently. Although this may sound like an esoteric problem, the ability to factor large integers efficiently is the key to breaking one of the most secure cryptographic schemes currently in use—the RSA cryptosystem, invented by Ronald Rivest, Adi Shamir, and Leonard Adleman (the letters come from their last names) in 1978 (Rivest, 1978). Today, if you use a credit card to buy products over the Internet, there is a good chance that you will be using the RSA cryptosystem to protect your confidential transaction.

In this chapter, we describe how the RSA cryptosystem works and how to break the RSA code using first a classical computer and then a quantum computer.

The Art of Concealment

When sensitive information (such as military plans, banking transactions, or confidential messages) is exchanged, people routinely scramble, or "encrypt," the information using a secret code. Such messages are secure as long as unscrambling, or "decrypting," the message is computationally intractable.

The use of coded messages is hardly new. The first known coded text is a small piece of Babylonian cuneiform writing dating from around 1500 B.C. It describes a secret method for making a glaze for pottery. By recording the instructions in a code, the potter could protect his secrets from the eyes of jealous competitors.

The ancient Greeks used a cryptographic scheme known as the *scytale*. It consisted of long, thin strip of cloth wrapped around a stick on which a message was written in vertical columns. When the cloth was unwound and removed from the stick, the letters that made up the message appeared to be scrambled. However, the message could be decoded by a recipient who possessed a stick that was shaped identically to the original stick on which the cloth was wound when the secret message was written.

The Roman emperor Julius Caesar is known to have used a transposition code in which each character in a message was displaced four characters ahead in the alphabet. Thus the message "Brutus might betray me" would have been encrypted as the text string "Fvyxyw qmklx fixvec qg."

In more modern times, machines have been developed to encode and decode messages using sophisticated new codes. The Enigma machine was invented around 1918 by Arthur Scherbius, who promoted its use for secure banking communications. It was described in a patent in 1919 and was adopted by the German navy in 1926, the German army in 1928, and the German air force in 1935. It was withdrawn from public use after the German navy adopted it, but it had already become known worldwide, throughout the cryptographic community. It was certainly known to the British government; there is a record of a meeting at which it was determined that the Enigma machine was not suitable for military use.

The Germans thought otherwise, however. In World War II, they used the Enigma machine and an even more elaborate Lorenz machine to pass orders between the German High Command and officers in the field. The Enigma code was used for communications with German submarines called U-boats, and the Lorenz code, known as the Fish code to the British code breakers, was used for communications among the higher echelons of the German army.

The breaking of Enigma- and Lorenz-encrypted coded messages quickly became a military priority. The British and the Americans both established code-breaking centers staffed by some of the best minds of their generation. In Britain, the center was located within a manor house known as Bletchley Park. Brian Oakely, former member of the Bletchley Park team and now curator of the Bletchley Park Trust recalls that these elite teams were not composed exclusively of mathematicians but included bright people with a knack for solving puzzles (Oakley, 1998). For example, at Bletchley Park, Dilly Knox, an eminent Greek scholar, unscrambled Enigma-encoded messages of the Italian navy once or twice during the Spanish Civil War. Bill Tute, at the time a biologist, reconstructed in 1942 and 1943 a Lorenz machine that enabled him to tackle some of the Fish codes.

Despite the skill of these individuals, the need to use a machine to break the codes quickly became apparent. A machine that reads in a coded message and unscrambles it by applying some systematic algorithm is nothing other than a Turing machine, so it is not surprising that Alan Turing was enlisted into the British code-breaking effort during World War II.

Turing joined the Code and Cipher School at Bletchley Park. The first code-breaking machines, called Turing Bombes, were derived from simpler Polish code-breaking machines of the 1930s. Three Polish mathematicians working for the Polish government first cracked the Enigma-encoded messages of the German military in 1930. However, the ability was lost when the Germans began to use more sophisticated machines. The Turing Bombes were electro-mechanical contraptions built to the design of Alan Turing with contributions from others, such as Gordon Welchman, a Cambridge mathematician. They employed relays, parts of tabulating machines, and special rotors and were programmed by means of plug boards. Weighing about a ton each, they started breaking the Enigma codes in the summer 1940. In all, 210 Turing Bombes were built in the United Kingdom and 200 in the United States by the end of the war. They were extremely effective. Breaking some keys virtually every day, they broke a total of 186 different keys during the war. By 1943 the British and Americans were reading an average of 2000–3000 secret German messages a day. Turing Bombes took from 14 to 55 hours to break each Enigma-encoded message—less when the Germans were found to be reusing certain keys.

Speed was crucial: A message had to be decrypted before the event threatened in that message took place. The old Turing Bombes were simply too slow to get the job done in time. Consequently, British Intelligence commissioned the construction of an electronic code-breaking machine based on vacuum tubes, which are much faster than electro-mechanical relays. At the time, there was no computer industry as such, so the British had to make do with the best they had—the Post Office, which also controlled the telephone system. The postal service simply had more experience than anyone else in the use of vacuum tube technology.

The first vacuum tube code-breaking machine used at Bletchley Park was called the Heath–Robinson (known as a Rube Goldberg device in America) because of its outlandish design. The Heath–Robinson used two tape loops that had to be kept synchronized and that broke frequently. They were never used operationally, but they demonstrated that a vacuum-tube–based "computer" could break the Fish codes produced by the Lorenz machines. The real break-through for unscrambling the Fish codes came with the introduction of the "Colossus" computer, which employed some 3500 vacuum tubes. In all, 13 Colossus machines were built. Typically, they could crack a Fish code in one or two days (such a machine once succeeded in doing so in 20 minutes).

On November 12, 1940, at the height of World War II, British military intelligence intercepted an Enigma-encoded message from the German High Command to the commander of the Luftwaffe. The Germans had no idea that the Allies could break the Enigma code and were remarkably explicit in communicating their true intentions in their coded messages. This particular message outlined a plan for a bombing raid November 14 on the city of Coventry, an industrial center in the middle of England. Unfortunately, the British code breakers misinterpreted the reference to Coventry and instead warned Winston Churchill, the British prime minister, that the raid was probably to be on London, but that there was a possibility it could be on Coventry or Birmingham. At 3:00 P.M. on November 14, the Radio Countermeasures team, a group independent of the cryptographic team, detected intersecting directional radio beams over the city of Coventry, which confirmed that Coventry was indeed the target. It is not entirely clear what happened next. It is most likely that in the fog of war, the message was not relayed to a high-enough authority to be acted on effectively. It is less likely, though possible, that Winston Churchill or other senior British officers deliberately failed to alert the people of Coventry of the impending raid so that the Germans would not suspect that the Enigma code had been broken. If Churchill *had* known the target for the air raid, he would probably not have announced it publicly but would instead have beefed up the anti-aircraft defenses around Coventry. The ack-ack defenses *were* in fact increased around Coventry on the night of the November 14, but this might simply have been in response to the possibility of the raid being on Coventry. Tragically, the ensuing air raid on Coventry killed hundreds of people, though it is likely that many more lives were saved with the knowledge gleaned from intercepted German messages in the months that followed.

Clearly, the ability to make and break secret codes is a crucial skill in the modern world. But just what kinds of codes are possible? How do you make a coded message in such a way that it is hard for an unintended recipient to unscramble it?

Encryption Schemes

There are numerous ways to encrypt a message. Whatever method is employed, the idea is to scramble a message in such a way as to make it unintelligible to potential adversaries yet understandable to intended recipients. Typically, this involves converting the text of the message, traditionally called the *plaintext,* into a sequence of numbers and then mapping these numbers into other numbers by applying some "encryption algorithm" to the numerical encoding of the plaintext. When the encrypted message arrives, the recipient must decrypt it back into the sequence of numbers and reconvert these numbers to the plaintext, that is, the original (secret) message. As a shorthand way of describing this process, given a plaintext message M, this message is converted into an encrypted message E using $E = \text{Encrypt}[\text{MessageToIntegers}[M]]$, where the inner function MessageToIntegers is computed first, creating an encoding of a plaintext message in a set of "message integers," and then the encryption function, Encrypt, is applied to these message integers. This process returns E, the encryption of M.

Before secure messages can be exchanged, the parties who wish to communicate must agree on the "alphabet" of symbols from which future messages will be composed. Typically, such an alphabet contains more than merely lower-case letters. It may also include punctuation marks, upper-case letters, numbers, and parentheses. The exact composition of the alphabet is not important. All that matters is that both parties agree on the set, and alphabetic ordering, of symbols to be used.

Let's look at an alphabet sufficient for simple communications. This particular alphabet, which we have called $Alphabet, contains 76 distinct symbols numbered from 0 to 75 (including "blank space" symbol).

$$\$\text{Alphabet} = \{\text{"a", "b", "c", "d", "e", "f", "g", "h", "i", "j",}$$
$$\text{"k", "l", "m", "n", "o", "p", "q", "r", "s", "t",}$$
$$\text{"u", "v", "w", "x", "y", "z", "A", "B", "C", "D",}$$
$$\text{"E", "F", "G", "H", "I", "J", "K", "L", "M", "N",}$$
$$\text{"O", "P", "Q", "R", "S", "T", "U", "V", "W",}$$
$$\text{"X", "Y", "Z", "1", "2", "3", "4", "5", "6", "7",}$$
$$\text{"8", "9", "0", "!", "?", " ", ",", ";", ":", "(", ")",}$$
$$\text{"\{", "\}", "[", "]", " ", "'"\}}$$

Given our 76-character alphabet, it is straightforward to define the MessageToIntegers function. If the alphabet contains ℓ distinct symbols, we can convert the message to a sequence of integers by numbering the symbols in the alphabet from 0 to $\ell - 1$ and then substituting the appropriate integer for successive symbols that appear in the message. Thus the secret message "My PIN number is 1234!" would, using $Alphabet, be converted into the sequence of integers {38, 24, 74, 41, 34, 39, 74, 13, 20, 12, 1, 4, 17, 74, 8, 18, 74, 52, 53, 54, 55, 62}.

Likewise, to decode a message given a sequence of such integers, we simply have to invert the substitution operation. That is, given the string of numbers above, we replace each integer with the corresponding symbol from $Alphabet to arrive at "My PIN number is 1234!"

Simple substitution alone, however, does not lead to a secure message. Substitution codes can be broken easily if the intruder knows the frequency with which the different symbols in the language in which the message is written occur across a representative sample of texts. For example, in written English, different letters occur with the frequency distribution shown in Figure 4.1. Knowledge of this distribution allows an adversary to guess large parts of a substitution code.

To ensure that a message will be difficult for an adversary to read, we need to do more than merely substitute symbols; we need to *encrypt*—that is, scramble up—the message in some way. For the code to be an *effective* method of communications, both the encryption and the decryption procedure must be reasonably *fast* computations. Unfortunately, this is unlikely to make for a secure communication, because if decryption is fast, an adversary may be able to guess the decryption rule through trial and error.

Trapdoor Functions

For the purpose of establishing *secure* communications, the encryption and decryption procedures need to possess another property in

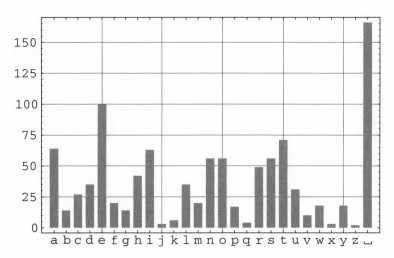

Figure 4.1   The average number of times each letter occurs in 1000 consecutive characters of written English. The blank-space symbol (shown on the far right) is the character that occurs between words and is denoted as ␣ in the figure.

addition to being fast to compute for the sender and the legitimate recipients of a secret message. The person sending a message would like to be able to encrypt the plaintext efficiently but also arrange for it to be almost impossible for an adversary to decrypt it in any reasonable length of time. One approach is to use a mathematical procedure that is easy to compute but hard to invert. There are many mathematical procedures that have this property of being easy to compute but hard to invert. A good example is multiplication, which is inverted by factorization. A prime number is a number that can be divided (with zero remainder) only by 1 and itself. For example, 2, 3, 5, 7, 11, and 13 are all prime numbers, but 9 is not because it is divisible by 3. Given two large prime numbers p and q, it is very easy to compute their product $n = p \times q$. For example, if $p = 15485863$ and $q = 15485867$, it is easy to determine that their product is 239812014798221.

On the other hand, being supplied with a large integer $n = 239812014798221$ the task of finding two integers p and q such that $p \times q = n$ is a *much* more difficult problem. In fact, the vast disparity between the ease of multiplying numbers and the difficulty of factoring them enables us to regard multiplication as a "one-way" function: easy to compute but hard to invert. One-way functions are used extensively in cryptography.

Unfortunately, there is a slight drawback. If we rely purely on one-way functions, decrypting the message is as hard for an intended recipient as it is for an adversary. This calls for a minor refinement of the idea, the notion of a "trapdoor" function.

A trapdoor function is some mathematical procedure that is easy to compute but very hard to invent *unless you have access to a special "key."* The most powerful modern cryptographic techniques for making unbreakable codes use the concept of a trapdoor function.

To encrypt a message using a trapdoor function with a key K, the sender computes $E = \text{Encrypt}[\text{MessageToIntegers}[M], K]$. Thus the encryption step involves both the number obtained by merely translating the plaintext according to some alphabetic substitution rule and a new special number, K, called a key. Upon receiving an encrypted message, a recipient unscrambles it by computing $M = \text{IntegersToMessage}[\text{Decrypt}[E], K]$.

Provided that the parties know the key, K, encrypting the message is easy for the sender and decrypting it is easy for the recipient. However, an adversary who intercepts the encrypted message but does not know the secret key will find it effectively impossible to unscramble the message.

One-Time Pads

The "one-time pad" (OTP) is a cryptosystem based on a trapdoor function. This particular cryptosystem is so secure—in fact it is *guaranteed* to be secure—that it is rumored to be used for communicating diplomatic information between Washington and Moscow (Welsh, 1988). Yet despite its high level of security, the one-time pad cryptosystem is quite simple.

Suppose two parties, Alice and Bob, wish to send each other secure messages. Before any secure communications commence, Alice and Bob must meet covertly to create a huge number of secret keys in the form of random integers picked uniformly in the range 0 to $\ell - 1$, where ℓ is the number of symbols in the alphabet. Typically, these keys are printed in a booklet or "pad," which is where the word *pad* in the name *one-time pad* originates. For example, a key pad containing 5 pages with 25 keys per page might be as follows (here each row of numbers represents a page of keys for the key pad):

$Pad = \{\{37, 69, 40, 19, 17, 65, 34, 26, 62, 32, 29, 57, 31, 27, 56, 53, 36, 15, 52, 72, 7, 48, 48, 19, 41\},$
$\{05, 75, 70, 18, 56, 15, 15, 09, 44, 41, 00, 72, 26, 31, 20, 37, 36, 23, 41, 19, 38, 63, 1, 68, 18\},$
$\{30, 57, 26, 33, 36, 75, 52, 16, 01, 70, 48, 14, 42, 23, 15, 20, 28, 45, 34, 51, 55, 37, 06, 8, 66\},$
$\{32, 73, 68, 22, 00, 70, 57, 00, 9, 24, 42, 26, 32, 45, 46, 47, 14, 35, 10, 59, 35, 24, 62, 66, 13\},$
$\{54, 36, 71, 01, 28, 23, 26, 39, 04, 67, 23, 33, 07, 09, 38, 37, 10, 32, 5, 64, 73, 63, 32, 20, 68\}\}$

Alice and Bob guard each of their copies of the key pad and return home. Now the parties are ready to communicate secret messages using the one-time pad cryptosystem. The steps that must be taken for Alice to send a secure message to Bob are listed in Table 4.1.

Here is an example of a one-time pad cryptosystem. We will use the 76-character alphabet $Alphabet and the key pad $Pad that we created earlier.

Suppose Alice wants to send Bob the secret message "My PIN number is 1234!" They have already agreed to exchange messages composed only of symbols from $Alphabet, and each has a copy of $Pad.

Given Alice's choice of which page to use from the key pad, the key pad itself, and the common alphabet, she can create a one-time pad encryption of a message. For example, if Alice uses page 3 of her key pad, the message "My PIN number is 1234!" becomes $E = \{68, 5, 24, 74, 70, 38, 50, 29, 21, 6, 49, 18, 59, 21, 23, 38, 26, 21,$

Table 4.1
One-time pad cryptosystem.

---

1. Alice converts her message into a sequence of integers. Thus the original plaintext, consisting of N symbols, becomes the N message integers $M = \{m_1, m_2, \ldots, m_N\}$.

2. Next Alice chooses a page, P, of keys from her copy of the key pad that she shares with Bob. This provides a supply of random integers $\{k_1, k_2, \ldots\}$. Alice only needs to use the first N such keys to encrypt M.

3. To perform the encryption, Alice computes a sequence of N encrypted integers $E = \{e_1, e_2, \ldots, e_n\}$ by applying the rule $e_i = m_i + k_i \ (\text{mod } \ell)$; that is the ith encrypted integer is obtained by adding the ith message integer to the ith key on page P of the key pad, dividing the result of ℓ, and keeping the remainder.

4. Alice sends Bob (E, P), the encrypted message E and page number P of the keys that she used.

5. Upon receiving them, Bob looks up the keys on page P of his copy of the key pad and finds the keys that Alice used, $\{k_1, k_2, \ldots\}$. Using the encrypted message $E = \{e_1, e_2, \ldots, e_n\}$, Bob reconstructs the message integers $M = \{m_1, m_2, \ldots, m_N\}$ by computing $m_i = e_i - k_i + \ell(\text{mod} \ell)$.

6. Finally, Bob converts M back into the original message using the Integers-ToMessage operation, which is the inverse of the MessgeToIntegers operation.

11, 29, 34, 23 }. Bob can decrypt Alice's coded message using page 3 of the key pad to get $M = $ My PIN number is 1234!

The OTP is guaranteed to be secure because there is a set of keys that will convert any encrypted message to any possible decrypt. So you cannot gain anything by "guessing" keys.

Unfortunately, one-time pads consume vast numbers of keys and require the sender and receiver to conspire to exchange key sets covertly before sending each other secure messages. Moreover, the integrity of the entire cryptosystem rests on maintaining the secrecy of the key pads. Should one of the key pads ever fall into the wrong hands, an adversary could decrypt the messages easily. These factors make one-time pads of limited utility. A more practical scheme requires that there be a way to *distribute* the keys, in a secure fashion, without the sender and recipient having to meet face to face.

Public Key Cryptography

The RSA system (Rivest, 1978), is a cryptosystem that solves the *key distribution* problem. Unlike the one-time pad scheme, in RSA the sender and recipient do not need to meet beforehand to exchange secret keys. Instead, the sender and receive use *different* keys to encrypt and decrypt a message. This makes it significantly more practicable than the one-time pad scheme. In fact, today most secure communications are based on the RSA cryptosystem.

The basic idea is as follows. A person such as Alice who wishes to *receive* secret messages using RSA creates her own pair of keys, consisting of a "public key" and a "private key." She makes the public key known but keeps the private key hidden. When someone wants to send her a secret message, the sender Bob, Charles, or Dillon obtains the public key of the intended recipient and uses it to encrypt his message. Upon receiving the scrambled message, Alice uses her private key to decrypt the message. The trick is to understand how the public key and the private key need to be related to make the scheme work in an efficient yet secure fashion.

In an *efficient* cryptographic scheme, it must be easy for a sender to compute E, the encryption of the plaintext message M, given the public key $\$PublicKey$. In other words, the computation $E =$ Encrypt[MessgeToIntegers[M], $\$PublicKey$] must be simple. Moreover, it must also be easy for the intended recipient to decrypt an encrypted message, given the private key $\$PrivateKey$. That is, the computation $M =$ IntegersToMessge[Decrypt[E, $\$PrivateKey$] must be simple too. Furthermore, it must be easy to generate the pairs of public and private keys.

In a *secure* cryptographic scheme, it must be extremely difficult to determine the message M, given only knowledge of E and the public key $\$PublicKey$. Also, it must be extremely difficult to guess the correct key pair.

Such a dual-key scheme is called a "public key cryptosystem" (Welsh, 1988). It is possible to have different cryptosystems by choosing different mathematical functions for creating the key pairs and the encryption and decryption procedures.

The RSA system is just such a cryptosystem. It relies on the presumed difficulty of factoring large integers. Suppose Alice wants to receive secret messages from other people. To create a public key/private key pair, Alice picks two large prime numbers, p and q, and computes their product, $n = pq$. She then finds two special integers,

d and e, that are related to p and q. The integer d can be chosen to be any integer such that the largest integer that divides both d and $(p - 1) \times (q - 1)$ exactly (i.e., with zero remainder) is 1. When this is the case, d is said to be co-prime to $(p - 1) \times (q - 1)$. The integer e is picked in such a way that the remainder after dividing ed by $(p - 1) \times (q - 1)$ is 1. When this relationships holds, e is said to be the modular inverse of d. Alice uses these special integers to create a public key consisting of the pair of numbers $\{e, n\}$ and a private key consisting of the pair of numbers $\{d, n\}$. Alice broadcasts her public key but keeps her private key hidden. By broadcasting her public key, Alice makes it possible for anyone to send her a secret message.

Now suppose Bob wishes to send Alice a secret message. Even though Bob and Alice have not conspired beforehand to exchange keys, Bob can still send a message to Alice that only she can unscramble. To do so, Bob looks up the public key that Alice has advertised and represents his text message M_{text} as a sequence of integers in the range 1 to n. Let us call these message integers M_{integers}. Now Bob creates his encrypted message E by applying the rule

$$E_i = M_i^e \bmod n$$

(i.e., raise the ith message integer to the power e, divide the result by n and keep the remainder) for each of the integers M_i in the list of message integers M_{integers}.

Upon receipt of these integers, Alice decrypts the message using the rule

$$M_i = E_i^d \bmod n$$

The final step is then to reconvert the message integers to the corresponding text characters. Thus the RSA cryptosystem can be summarized as in Table 4.2.

For example, suppose Alice wants to set up a public key/private key pair based on a 20-digit integer. She could proceed as follows: Alice picks $p = 6257493337$ and $q = 6356046119$. Their product n is $pq = 39772916239307209103$. Alice picks any large integer d that is co-prime to $(p - 1)(q - 1)$, such as $d = 5380958597982080231$. Given d, p, and q, Alice can determine e, the modular inverse of d, from the rule $ed = 1 \bmod (p - 1)(q - 1)$. Hence Alice finds $e = 34928543677329462263$. Thus Alice has constructed the public key $\{e, n\} = \{34928543677329462263, 39772916239307209103\}$ and the private key $\{d, n\} = \{5380958597982080231, 39772916239307209103\}$.

Table 4.2
Summary of the RSA public key cryptosystem.

1. Find two large primes p and q and compute their product $n = pq$.

2. Find an integer d that is co-prime to $(p - 1)(q - 1)$.

3. Compute e from $ed \equiv 1 \bmod (p - 1)(q - 1)$.

4. Broadcast the public key—that is, the pair of numbers (e, n).

5. Represent the message to be transmitted, M_{text}, say, as a sequence of integers $\{M_i\}$ each in the range 1 to n.

6. Encrypt each M_i using the public key by applying the rule

$$E_i = M_i^e \bmod n$$

7. The receiver decrypts the message using the rule

$$M_i = E_i^d \bmod n$$

8. Reconvert the $\{M_i\}$ to the original message M_{text}.

Next Alice broadcasts her public key, *$PublicKey*, {3492854 3677329462263, 39772916239307209103}, but keeps her private key (needed for decrypting the message), *$PrivateKey*, {53809585 97982080231, 39772916239307209103}, secret. Hence the public key is available to everyone, but the private key is known only to Alice. If someone, say Bob, wants to send Alice an encrypted message that only Alice can decode, he simply looks up Alice's public key, converts the text message to a sequence of integers, and encrypts each of the message integers using the rule $E_i = M_i^e \bmod n$ as previously explained. Using the aforementioned *$PublicKey*, Bob can encrypt the message

"Sell all shares of Netscape now!"

as the following sequence of integers: 3210343086664497793, 25354410350102000730, 2018537172583222694, 253544103501 02000730, 2028580298523594573, 19787911989951759145, 673 5419423900161318, 37836439796525544215, 176251336663283 50954, 2935932729727566345, 9984261459086080108, 1281853 0531221364531, 2846177880364009532, 77850787591353508564, 2330216654924735484, 13268812316670558622.

This is the message—the sequence of integers—that is sent over the public channel. To decode the message, the recipient, Alice, uses her unique knowledge of her private key $\{d, n\}$. Alice first unscrambles each cipher integer using the rule $M_i = E_i^d \bmod n$ and then maps the message integers back into letters, symbols and numbers.

Alice and Bob have succeeded in encrypting and decrypting a secret message using the RSA protocol.

What makes RSA so useful is not merely the fact that there is an algorithm by which messages can be encrypted and decrypted but rather that the algorithm can be computed efficiently. Speed is vital if a cryptosystem is to provide a viable means of secure communication. Fortunately, each step in the RSA procedure can be done quickly. This may surprise you, because it is not immediately obvious that the calculations needed to find the pair of large prime numbers p and q and the special integers d and e can all be done efficiently. However, it turns out that p, q, d, and e can all be found quickly (Welsh, 1988). Thus every step in the RSA procedure can be computed efficiently, and this makes it a viable cryptosystem overall.

Does the ease of the computations underlying RSA mean that RSA is vulnerable to attack? To answer this, let us take a look at what an adversary would have to do to crack RSA-encoded messages.

Code Breaking on a Classical Computer

In order to break the RSA code, an adversary would need to find the private key $\{d, n\}$, given knowledge of the public key $\{e, n\}$. Once d is known, an intercepted encrypted message C can be decrypted by computing $C^d \bmod n$ just as the legitimate recipient would do. But how is an adversary to find the private key, given knowledge of the public key?

An adversary knows that the private key consists of the pair of numbers $\{d, n\}$, and the number n is disclosed in the public key $\{e, n\}$, so all the adversary must do is find the number d, given knowledge of the numbers e and n. A good way to start is to ask how d is related to e and n. Well, we saw before that the number e is the "modular inverse" of d. This means that e and d are related via the equation $ed = 1(\bmod (p-1)(q-1))$; that is, d is the number such that e times d divided by $(p-1)$ times $(q-1)$ leaves a remainder of 1. Thus if the adversary knows the three numbers e, p, and q, he can easily determine d and hence obtain the private key. Finding e is easy; like n, it is disclosed in the public key, $\{e, n\}$. Moreover, it is no secret that the numbers p and q are prime numbers whose product is equal to n—that is, $n = p \times q$. You might remember that the RSA encryption procedure begins by picking p and q. However, the mathematical operation that yields p and q, given n, is called factoring the number n. Therefore, an adversary's ability to break the RSA code all boils down to whether he can factor the number n in a reasonable length of time. Breaking RSA is difficult only if factoring large composite integers is also difficult.

Fortunately for those who want to exchange confidential messages securely, the computing time needed to factor an integer rises sharply with the "size" of the integer. You can think of the size of an integer as the number of bits needed to specify the integer. For example, the size of the number 15 is 4 bits because 15 in base-10 notation is represented as 1111 in base-2 (binary) notation which takes four bits. To date, the best-known classical algorithm for factoring large integers is the "Number Field Sieve" (Lenstra, 1993). This has a running time that grows faster than polynomially but slower than exponentially in the "size" of the number being factored. Accordingly, it is known as a super-polynomial-time (alternatively, as subexponential-time) algorithm. Mathematically, if the number being factored requires L bits to describe it, the best classical factoring algorithm has a running time that grows as $\exp(cL^{1/3}\log(L)^{2/3})$ where c is a constant. This is faster than any polynomial in L (i.e., any growth rate of the form L^N) and yet slower than any exponential in L (i.e., N^L). A computation whose running time scales superpolynomially with the size of the problem is generally regarded as being effectively intractable. (See Figure 4.2.)

Thus the security of the RSA cryptosystem rests on the presumption that factoring large integers is computationally intractable—practically impossible for large enough cases. We say "presumption" because no one has yet *proved* that factoring is a truly intractable problem. There is a possibility that someone may discover a polynomial-time classical algorithm. But so far, no one has. Umesh Vazirani, a computer scientist at the University of California Berkeley, has characterized the difficulty of factoring a 2000-digit number. Vazirani states (Vazirani, 1994), "It's not just the case that all the computers in the world today would be unable to factor that number. It's really much more dramatic. ... Even if you imagine that every particle in the Universe was a (classical) computer and

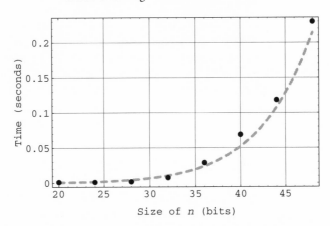

Figure 4.2 The increase in time needed to factor an integer on a classical computer as a function of the "size" of the integer. Note that the curve rises more steeply as larger integers are considered. This means that factoring large integers is *very* difficult.

was computing at full speed for the entire life of the Universe, that would be insufficient to factor that number."

Messages encrypted using the RSA scheme with a key containing 2000 digits should be quite impossible to break. Even messages encrypted using a 400-digit key are well beyond the decryption capabilities of modern computers. Thus anyone who wants to make an encrypted message secure need only pick a large enough value for *n* in the public key that such a number could not be factored within a reasonable time.

How big does *n* have to be, given the current state of (classical computer technology? When Rivest, Shamir, and Adleman first devised their cryptosystem back in 1977, they challenged anyone to factor a particular 129-digit number that came to be known as RSA-129. The challenge stood unmet for 17 years. Over that time period, computers became about 2000 times faster. Even then, RSA-129 was factored only after a Herculean effort involving a network of some 1600 computers.

Given the difficulty of factoring a 129-digit number, it is easy to see that basing an RSA cryptosystem on a number that involved, say, 150 digits would make any encrypted messages utterly secure for the foreseeable future. Secure, that is, with respect to the capabilities of hordes of classical computers. But what about quantum computers? Could a quantum computer factor large numbers fast enough to break the supposedly invulnerable RSA cryptosystem?

Code Breaking on a Quantum Computer

By 1994, a few scientists had recognized that quantum computers changed the rules of computational complexity theory, the study of the efficiency with which computers can perform computational tasks. Richard Feynman had speculated that quantum computers could simulate quantum systems more efficiently than any classical computer. David Deutsch and Richard Jozsa had shown that quantum computers could answer various questions concerning whether a function is even (i.e., always gives the same output) or is balanced (i.e., gives one output on half of the inputs and another output on the other half of the inputs). Various other separations between the behavior of quantum computers and that of classical computers had been found by Gilles Brassard, Andre Berthiaume, Umesh Vazirani, and others, but they were all rather abstract and not of much obvious applicability (see Chapter 3).

In 1994, however, Dan Simon showed that a quantum computer could obtain a sample from the Fourier transform of a function faster than any classical computer (Simon, 1994). This last result did not attract much attention at first, but it was to prove pivotal in the development of quantum computing.

Any mathematical function can be described as a weighted sum of certain "basis" or "elementary building block" functions such as sines and cosines, $\sin(x)$, $\sin(2x)$, ..., and $\cos(x)$, $\cos(2x)$, ... You might find this surprising because the graphs of sine and cosine are wavy, with the distances between successive crests and troughs fixed, whereas the graphs of other functions are not always so smooth and regular. For example, the dark curve in Figure 4.3 shows the graph of the function $y(x) = 1 - \exp(-3([x/2] - \lfloor x/2 \rfloor))$. Here the term $\exp(\cdot)$ denotes the exponential function and the term $\lfloor x/2 \rfloor$ is shorthand notation for the greatest integer less than or equal to $x/2$. As you can see the graph repeats itself periodically in a regular way, but it suffers from sharp kinks which you never find in sines and cosines.

Remarkably, the graph of $y(x)$ can be approximated as a sum of smoothly varying sines and cosines. This sum is plotted as the light gray curve in Figure 4.3. The actual form of the approximation is a sum involving just three types of sines and cosines:

$$y(x) \approx .683 - .118\cos(\pi x) - .034\cos(2\pi x) - .016\cos(3\pi x)$$
$$- .246\sin(\pi x) - .143\sin(2\pi x) - .098\sin(3\pi x)$$

As you might imagine, by allowing more sines and cosines to contribute to the sum the approximation can be made better and

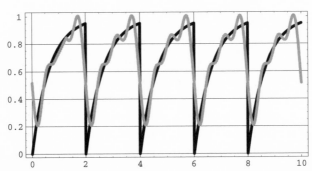

Figure 4.3 Graph of a function (dark) together with its Fourier series approximation (light).

better. This tells us that the sines and cosines provide a family of building block functions whose sums can be used to describe less regular functions and signals. In many scientific and engineering applications it turns out to be useful to take a signal and split it up into its principal sine and cosine components. The Fourier transform of a function is the mathematical operation that translates the original function into this equivalent sum of sine and cosine functions. If the Fourier transform of the original function is very spiky, it means that the function consists of just a few sines and cosines with precisely defined periods. Dan Simon arranged for a quantum computation to cause the state of a quantum memory register to evolve into a superposition representing the Fourier transform. By reading this memory register, we would most likely obtain a result corresponding to where the probability amplitude was most highly concentrated—that is, where the Fourier transform is most strongly spiked. Hence, a quantum measurement returns a *sample* from the Fourier transform, and this sample gives us some information about the periodic sine and cosine functions out of which our original function is built.

Peter Shor, a computer scientist at AT&T Research, was looking for a way to make a quantum computer solve a significant computational problem rather than the contrived problems that had been demonstrated so far. He realized that if he could relate the problem of finding the factors of a large composite integer to that of finding the period of a function, then he could use a technique akin to Dan Simon's for sampling from the Fourier transform. This insight led to the first "killer ap" for quantum computing when, in 1994, Peter Shor showed how a quantum computer could be used to factor a large integer super-efficiently (Shor, 1994). This was big news. Immediately the interest of security agencies and banks was piqued. If a quantum computer could factor an integer efficiently, it could also break the RSA cryptosystem. Suddenly, all kinds of secret information became vulnerable.

Let's take a look at the trick that relates finding the period of a function to finding the factors of a large composite integer.

A Trick from Number Theory

Shor's quantum algorithm for factoring relies on a result from number theory that relates the period of a particular periodic function to the factors of an integer.

Given an integer n (the number to be factored), think about a related function, $f_n(a) = x^a \bmod n$, where x is an integer chosen at random such that the largest number that divides both x and n is 1 and the mod function calculates the remainder after dividing x^a by n. In mathematical parlance, the fact that the largest integer that divides both x and n is 1 means that the "greatest common divisor" of x and n is 1—that is, $\gcd(x, n) = 1$—or, equivalently, x is "coprime" to n. Both the mod function and the gcd function can be computed efficiently (even classically), so obtaining the values of $f_n(a)$ for various inputs a is inexpensive computationally.

Why is this remotely interesting with respect to the problem of factoring n? It turns out that the new function we just constructed, $f_n(a)$, is periodic; that is, for successive inputs $a = 0, 1, 2, \ldots$, the output values $f_n(0), f_n(1), f_n(2), \ldots$ fall into a repeating pattern eventually. Different values of x give rise to different patterns. The number of values between repetitions of the pattern, for a particular value of x, is called the *period, r*. Once you know this period r, number theory predicts that there is a very good chance (though not absolute certainty) that the largest integer that divides both $x^{r/2} - 1$ and n will be a non-trivial divisor of n. In other words, $\gcd(x^{r/2} - 1, n)$ is likely to be a factor of n. A trivial divisor would be 1 or n. Moreover, there is an algorithm known from the ancient Greeks that enables us to compute the greatest common divisor of two numbers very efficiently. Thus, if the period r can be found efficiently, there is a good chance that the factors of n can be found efficiently.

Now this all sounds a bit abstract, so let's see how it works on a simple example. Suppose you want to factor the number $n = 15$. You need to pick a number x such that the largest integer dividing both x and n is 1. Let's say you choose $x = 8$. Next compute the value of the function $f_n(a) = x^a \bmod n$ with $x = 8$ and $n = 15$—that is, $f_{15}(a) = 8^a \bmod 15$—for $a = 0, 1, 2, \ldots$. You find that $f_{15}(0) = 1$, $f_{15}(1) = 8$, $f_{15}(2) = 4$, $f_{15}(3) = 2$, $f_{15}(4) = 1$, $f_{15}(5) = 8$, $f_{15}(6) = 4$, and so on. Sure enough, the pattern of outputs forms a periodic sequence, namely, 1, 8, 4, 2, 1, 8, 4, 2, 1, 8, 4, 2, 1,.... The period of this sequence (the number of steps between repetitions of the pattern) is $r = 4$. Finally, you compute the greatest common divisor of $x^{r/2} - 1$ and n and find that $\gcd(63, 15) = 3$. Indeed, the number 3 is a non-trivial divisor of the number 15. Once you have one divisor, it is easy to determine the other by division: $15/3 = 5$. The factors of 15, then, are 3 and 5.

The key point here is that the factors of an integer can be obtained by finding the period of a related function. Unfortunately, there is no known way to calculate the required period *efficiently* on any classical computer. However, Pete Shor, building on the work of Dan Simon, discovered a way to calculate the period efficiently on a quantum computer.

Shor's Algorithm for Factoring Integers on a Quantum Computer

How does Shor's algorithm work? Let us start off by giving the big picture and focus on the details later. Our goal is to find the period of the function $f_{x,n}(a) = x^a \bmod n$. To do so, we create a single quantum memory register that we partition into two parts called Register1 and Register2. Although the complete register consists of a chain of qubits, we are going to use a more compact notation for representing its contents. If Register1 is holding the (base-10) number a and Register2 is holding the (base-10) number b, then we represent the state of the complete register as $|a, b\rangle$.

Next we create, in Register1, a superposition of the integers $a = 0, 1, 2, 3, \ldots$ that represents all possible inputs to the function $f_{x,n}(a)$. Then we evaluate, in quantum parallel, $f_{x,n}(a)$ on this superposition and place the result in Register2. This creates a superposition, in Register2, of the function evaluations $f_{x,n}(a)$ in the time it takes to compute just one such function evaluation classically. Next we measure the state of Register2. This collapses the superposition stored in Register2, and we obtain some answer, say k. This means that there was some value of a such that $x^a \bmod n = k$.

However, this simple act of measuring the state of Register2 has a critical side effect on the state (i.e., the contents) of Register1. Remember that Register1 and Register2 are simply two parts of a single quantum register. Because Register1 and Register2 are entangled with one another, by observing Register2 we actually change the contents of Register1. This is one of the true miracles of quantum computing: the possibility of exploiting entanglement to bring about desired, correlated side effects during a quantum computation. After the measurement made on Register2, the state of Register1 becomes a superposition of just those values of a such that $x^a \bmod n = k$. The values now stored in Register1 are of the form $\{a, a + r, a + 2r, \ldots\}$; that is, they implicitly encode multiples of the sought-after period r. If only we could extract that period, we could use the trick from number theory to find the factors of n.

To expose the period, we compute the Fourier transform of the contents of Register1 and put the result back into Register1. As we have said, the Fourier transform of a function is spiky if the function implicitly harbors some periodicity. Register1 does indeed contain a function that has just such a hidden periodic behavior, so we hope that the Fourier transform will expose this periodicity.

Once we have performed the Fourier transform, Register1 contains a periodic function, peaked at multiples of 1 over the period, that is, $1/r$. Thus, if we now measure the state ("read the contents") of Register1, we are very likely to obtain a result that is close to some multiple of $1/r$. In particular, the result we read from Register1 will, with high probability, be the best integer approximation to a multiple of $2^\ell/r$, i.e., $x = k2^\ell/r$ for some integer k. Knowing the measured value of x and the size of Register1 (ℓ qubits) we can, if the integers k and r are coprime, then determine r by canceling $x/2^\ell$ down to an irreducible fraction and reading off its denominator. Because the probability of k and r being coprime is greater than $1/\log r$ for large r, this cancellation technique has a very good chance of revealing the sought-after period, r.

If we repeat this whole process just a few times, we will quickly obtain enough samples of integer multiples of $1/r$ to be able to determine the period r unambiguously.

Thus, Shor's quantum algorithm for finding the period of $f_{x,n}(a) = x^a \bmod n$ relies on quantum parallelism to create a superposition of values of the periodic function $f_{x,n}(a)$, relies on measurement to project out a periodic function in Register1, and relies on a Fourier transform to bring about the desired interference effect between solutions (integer multiples of $1/r$) and non-solutions (numbers that are not integer multiples of $1/r$).

Having found the period, we find the factors of n from $\gcd(x^{r/2} - 1, n)$ and $\gcd(x^{r/2} + 1, n)$. As in the purely classical case, the method will find non-trivial factors of n, provided that r is even and that $x^{r/2} \neq \pm 1 \mod n$. The steps in Shor's quantum algorithm are summarized in Table 4.3.

Table 4.3
Summary of Shor's quantum algorithm for factoring composite integers.

---

1. Pick a number q such that $2n^2 \leq q \leq 3n^2$.

2. Pick a random integer x whose greatest common divisor with n is 1.

3. Repeat steps (a) through (g) about $\log(q)$ times, *using the same random number x* each time.

   (a) Create a quantum memory register, and partition the qubits into two sets called Register1 and Register2.

   (b) Load Register1 with all integers in the range 0 to $q - 1$, and load Register2 with all zeroes.

   (c) Now compute, in quantum parallel, the function $x^a \mod n$ of each number a in Register1, and place the result in Register2.

   (d) Measure the state of Register2, obtaining some result k. This has the effect of projecting out the state of Register1 to be a superposition of just those values of a such that $x^a \mod n = k$.

   (e) Next compute the Fourier transform of the projected state in Register1.

   (f) Measure the state of Register1. This effectively samples from the Fourier transform and returns some number c' that is some multiple λ of q/r, where r is the desired period; that is, $c'/q \approx \lambda/r$ for some positive integer λ.

   (g) To determine the period r, we need to estimate λ. This is accomplished using a continued-fraction technique.

4. By repeating steps (a) through (g) we create a set of samples of the discrete Fourier transform in Register1. This gives samples of multiples of $1/r$ as $\lambda_1/r, \lambda_2/r, \lambda_3/r, \ldots$ for various integers λ_i. After a few repetitions of the algorithm, we have enough samples of the contents of Register1 to compute what λ_i must be and hence to guess r.

5. When r is known, the factors of n can be obtained from $\gcd(x^{r/2} - 1, n)$ and $\gcd(x^{r/2} + 1, n)$.

Example Trace of Shor's Algorithm

Here is an example of the use of Shor's algorithm in the task of factoring the number $n = 15$. In the following figures, we represent the contents of Register1 in the horizontal direction and the contents of Register2 in the vertical direction. Suppose we pick $n = 15$, $x = 13$, and $q = 243$. Initially we load both Register1 and Register2 with zeroes.

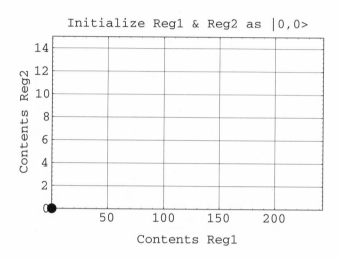

Next we load Register1 with a superposition of all possible inputs in the range 0 to $q - 1$. This is represented as a long horizontal strip in the figure.

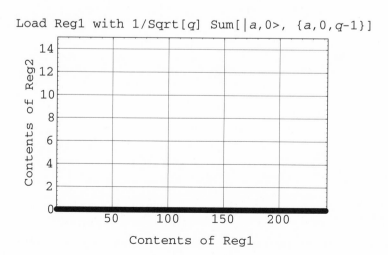

Next we compute, in quantum parallel, $f_{x,n}(a) = x^a \bmod n$ for each input a in Register1 and put the result in Register2. Register2 will then contain a superposition of states representing the sequence of values {1, 13, 4, 7, 1, 13, 4, 7, 1, 13, 4, 7, 1, 13, 4, 7, 1, 13, 4, 7, 1, 13, 4, 7, 1, 13, 4, 7, 1, 13, 4, 7, 1, 13, 4, 7, 1, 13, 4, 7, ...}.

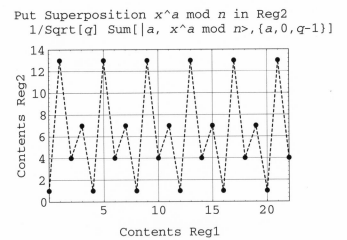

Put Superposition x^a mod n in Reg2
1/Sqrt[q] Sum[|a, x^a mod n>,{a,0,q-1}]

Next we measure the contents of Register2. Suppose we obtain the value 1 as our answer.

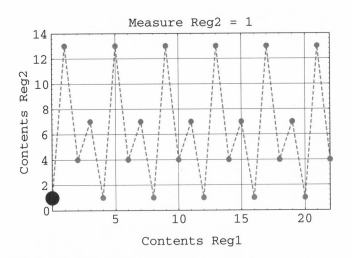

Measure Reg2 = 1

The measurement of Register2 projects Register1 into an equally weighted superposition of the states $|0\rangle + |4\rangle + |8\rangle + |12\rangle + |16\rangle + \cdots + |236\rangle + |240\rangle$.

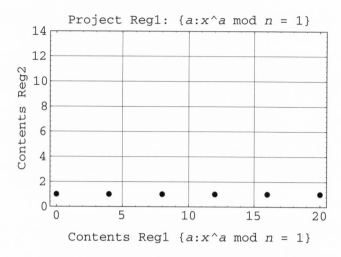

Next we compute the Fourier transform of the contents of Register1. This produces a superposition of the form $0.501|0\rangle + 0.002|1\rangle + \cdots$ in which the amplitudes are no longer equal. In fact, the Fourier transform has most of the amplitude concentrated in states that correspond to multiples of $1/r$, where r is the period sought.

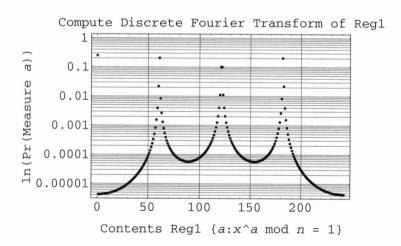

Upon observing the state of the Register1, we effectively sample from this Fourier transform. Suppose we obtain the result 61. We store this result and repeat the entire procedure.

Compute Discrete Fourier Transform of Reg1

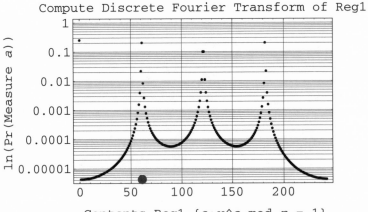

Contents Reg1 {a:x^a mod n = 1}

Multiple repetitions of the preceding steps might give us the following set of samples from the Fourier transform: {121, 61, 184, 182, 61, 122, 0, 121, 0, 121, 181, 61}.

Repeat Shor's Algm O($\ln(q)$) Times
Obtain Samples from DFT in Reg1

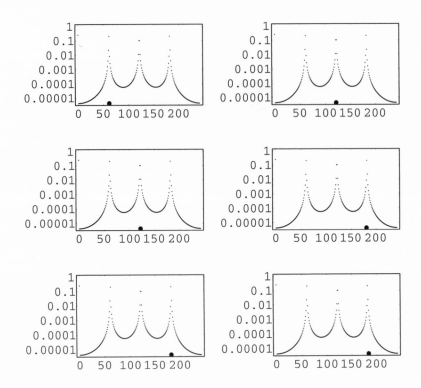

From these samples we can use a continued-fraction expansion, a technique from classical computer science, to deduce that the period sought, r, is 4. Hence we can obtain the factors of 15 by computing $\gcd(13^2 - 1, 15) = 3$ and $\gcd(13^2 + 1, 15) = 5$. Thus, the factors of 15 are 3 and 5.

It is important to realize that Shor's algorithm is probabilistic: It does not always return a non-trivial factor. For example, if we wanted to factor $n = 15$, but we picked $x = 14$, then the periodic sequence we would produce in Register2 would be 1, 14, 1, 14, 1, 14, 1, 14, 1, 14,. . . . Hence the period r in $f_{15,14}(a) = 14^a \mod 15$ is $r = 2$. Unfortunately, however, when we compute $\gcd(x^{r/2} - 1, n)$ and $\gcd(x^{r/2} + 1, n)$ we get a surprise: $\gcd(14 - 1, 15) = 1$ and $\gcd(14 + 1, 15) = 15$. Therefore, in this case Shor's algorithm yields only the trivial divisors of 15, namely, 1 and 15.

Summary

Shor's quantum algorithm has recently been combined with Grover's algorithm (see Chapter 3) to perform quantum counting—that is, to determine how many solutions a problem admits without revealing any of those solutions explicitly (Brassard, 1998). This in itself is useful because it enables us to generalize the Grover search procedure to the case in which the "haystack" might contain several equally acceptable "needles." Specifically, if the virtual database contains S sought-after items, any one of which is acceptable to us, then it takes only $O(\sqrt{N/S})$ steps to find one of the items using a quantum computer.

Shor's algorithm shows us that factoring is in P for a quantum computer. But no one has yet *proved* that factoring is not in P for a classical computer. It is therefore still possible that someone will find an efficient (polynomial-time) classical algorithm for factoring. Moreover, if scientists can find an efficient way to simulate a quantum computer on a classical computer, then this would immediately admit an efficient classical factoring algorithm that could be created merely by simulating a quantum computer running Shor's algorithm. However, it seems impossible to simulate a quantum computer efficiently on a classical computer. And indeed it may be impossible.

Currently, researchers are attempting to build a prototype quantum computer that will be capable of factoring a small integer. It is likely that we will see a quantum computer specialized to factor the number 15 by around the year 2000. Chapter 9 describes some of the efforts to build real quantum computers.

This chapter has shown that quantum computers can be more efficient than classical computers for at least one significant application, code breaking. In the next chapter we describe an even more spectacular advantage. We show how quantum computers can perform a task that cannot be performed, even in principle, by any classical machine. This capability arises because classical computers can compute *only* functions. Certain computational tasks, such as picking a random number, cannot be accomplished by computing any function. Classical computers therefore have to fake the performance of such tasks. Quantum computers, however, can perform these tasks directly because they can do more than mere function calculation.

Five

⤜⟫◆⟪⤛

The Crapshoot Universe

God does not play dice.
—Albert Einstein

Einstein, stop telling God what to do.
—Niels Bohr

In this chapter we explore the concept of randomness and how we can and must harness it in the service of computation. At one extreme, randomness is as mundane as the outcome of a coin toss or the roll of dice. At the other extreme, randomness reaches to the very core of the metaphysical profundities of quantum mechanics. Quantum physics, as it is usually taught and practiced, relies heavily on the concept of randomness. In particular, when a physical system that is in a superposition of states is observed, it is as though that superposed state collapses, as a result of this measurement, into one of the eigenstates. We cannot predict which of those states will appear as a result of our measurement. All we can do is give the probability of obtaining the various possible outcomes. The inability to predict the state into which a system will collapse upon being observed adds the element of randomness to quantum theory.

Whereas the previous chapter illustrated that a quantum computer can be more *efficient* than a classical computer in performing certain computations, in this chapter we shall see that a quantum computer is more *proficient* than a classical computer at certain important tasks, such as simulations. A classical computer can only *pretend* to generate randomness, whereas a quantum computer can *actually* do so.

This may seem like a minor point—hardly worth an entire chapter—but as we shall see, it has profound implications for the future of many important computer-based simulations that must make choices among possible outcomes.

The Concept of Randomness

What is true randomness? It is at best an elusive concept. To many people, randomness jumbles together the notions of unpredictability, disorder, chaos, and unintentionality. In other words, we think of randomness as referring to some phenomenon, such as a coin toss, for which we cannot predict the outcome with *certainty*. Note that randomness does not mean that we can say *nothing* about a phenomenon; we can in fact make extremely accurate predictions about what the *average* behavior will be over the course of many trials. For example, we know that there is a 50 percent chance of getting a head or a tail on any given toss of a fair coin. It is just that we cannot predict exactly which outcome will occur on a *particular* trial. But is this uncertainty of outcomes something that can be overcome with a greater knowledge of the system under study? Can we measure the dynamics of a coin toss exactly and then predict the correct outcome each time?

We can, according to classical physics. What would we need to know? We would need to know the exact, and we mean *exact*, initial conditions of the coin just before the toss, the angle of the coin, its weight, and the initial "flip" strength, and we would require detailed knowledge of the surface on which the coin will land. What no one realized in the days of classical physics was that there are not enough atoms—or, more precisely, bits—in the universe even to record a measurement with infinite precision. Thus, determinism in this sense is a lost cause.

Since Newton's time over 200 years ago, statisticians have made the concept of randomness quite precise. A hundred years ago physicists finally utilized these results when they realized that rather than deal with the details of complicated physical systems directly, they could use theories of randomness to model them. This led to the sub-field of physics known as statistical mechanics, which is concerned with the behavior of systems with a large number of components. Surprisingly, though, the deterministic theories of classical physics were never fully reconciled with the probabilistic theories used in thermodynamics and statistical mechanics. And this despite the fact that the statistical theories were supposed to be derivable from the deterministic theories.

From a computer science perspective, theories of randomness involve creating programs for generating sequences of random bits. Such bit sequences can be used to determine a sequence of events in the simulation of a complicated process that may have multiple possible outcomes only one of which is actually realized. The random sequence, then, is a proxy for a highly complicated and possibly non-computable process. A truly random sequence cannot be created by a program running on a classical deterministic computer such as a deterministic Turing machine, because the behavior of such a machine is predictable and thus not random.

But the lure of harnessing randomness remains because in it lies simplification. And the scientist looks for simplification wherever possible. For example, in modelling a coin toss we substitute all the equations of motion and properties of materials involved in the coin-flipping system and reduce them to a single random bit for each coin toss. To put it more bluntly, randomness is a proxy for our ignorance.

Unfortunately, things are not that easy. Although the theory of randomness may be precise, converting it into a usable procedure is not trivial. For instance, how do we know whether a particular procedure produces results that are "random enough" for a particular application? What appears sufficiently random from one perspective may not be very random at all from another.

To begin with, there is no such thing as *a* random number. For instance, 42 is not a random number by itself. It only makes sense to talk about processes for generating *sequences* of random numbers. Thus, . . . 23 42 14 . . . may be part of a random sequence, but then again it may not be (e.g., when it is part of the sequence 1 7 23 42 14 1 7 23 42 14 . . .). Thus we are faced with the question of whether a particular sequence of numbers has the *appearance* of being random.

Fortunately, there are statistical tests that can answer this question to a degree. To pass a test for randomness, the numbers in the sequence must follow some distribution, or pattern of occurrence. For example, the dot count of a single die throw forms a uniform distribution of integers, that is, the numbers 1, 2, 3, 4, 5, and 6 are equally likely to be obtained in any throw of the die. This means that if we throw the die 600 times, we should expect that, on average, each face of the die will appear 100 times. However, the sum of the dot count for two dice thrown together is not uniformly distributed but peaks at 7 and is less for a higher or lower number of dots. With even more dice thrown, the dot count distribution becomes even more sharply peaked. In general, if we add together many such independent, identically distributed random processes with finite averages and non-zero variances, then the sum of the outcomes of the processes converges to a bell-shaped curve (Grimmett, 1992). It is the emergence of such regularities from a myriad of simultaneous effects that makes the statistical viewpoint so effective.

Despite following some kind of distribution, the sequence of numbers generated should exhibit no discernible correlations. For example, if we were simulating repeated tosses of a fair coin and we obtained the sequence of outcomes H, T, H, T, H, T, H, T, H, T, H, T, H, T, H, T, H, T, H, T, H, T, . . ., this sequence would certainly pass the distribution check because the results are uniformly distributed between the two values H and T. However, it would not pass the correlation check; it seems as though the outcomes alternate systematically between H and T, so the next outcome of this sequence would be easy to predict correctly.

The challenge of randomness for computation is in the art of creating computer programs that simulate the generation of true random numbers. In other words, we need to devise algorithmic methods that generate sequences of numbers that pass both the distribution check and the correlation check. Unfortunately, as we shall see, even when a random-number generator passes such statistical tests, the sequence of numbers it generates may still not be random enough to serve as an approximation to a true random process.

Given the problems with identifying random numbers, we might be tempted to ask why we care about random numbers in the first place. Are they really so critical that we must have true randomness? Are there applications that depend critically on true randomness, or is the whole thing a mere mathematical muse?

Uses of Random Numbers

Even if we do know the exact equations that govern some dynamical system and the exact initial condition in which the system starts off, it can be a daunting, if not infeasible, computational task to use these equations for predicting the future of the system. For example, consider the enormous intricacy involved in describing the workings of the world economy. Consider further the incredible complexity of the billions of human brains that go into driving that world economy. With 100 million neurons firing 10 times per second per person, we have something like 60 exabits (60×10^{18} bits) per second determining the world economy, even ignoring unexpected events such as natural disasters. We cannot possibly hope to model this processes directly, yet it is of vital importance to every person on the planet. What can we do?

A statistical description—that is, a description involving probabilistic or random occurrences—is often more manageable, and more informative by virtue of its simplicity, than an exact description, particularly in cases involving thousands or even millions of components. In fact, the greater the number of possible outcomes, the better a statistical approach based on random numbers may perform.

Another use of random numbers occurs in certain types of games. In these games the optimal plays for winning constitute a purely random strategy. Any deviation from random behavior by one player—a bias—would allow the other player, over the course of many games, to obtain a consistent advantage.

We encountered important uses for random numbers in the last chapter. Recall the one-time pad cryptosystem. This cryptosystem relies on the sender and receiver each having a copy of a secret pad of "random" numbers, or "keys," by which they encrypt and decrypt secret messages. The one-time pad cryptosystem is provably secure, provided that the key pads are known only to the communicating parties.

In fact, during World War II, the Soviets routinely used a one-time pad cryptosystem, called the Vernam cipher, for their diplomatic communications. Unfortunately, they sent so many secret messages that they consumed all the random keys they had pre-computed in their key pads. Consequently, instead of deleting used pages of keys from their key pads, they began to reuse them. But reusing the keys caused a discernible pattern to emerge because the keys were no longer random—they repeated themselves after a time. This enabled cryptanalysts in the West to break the Soviet codes. Eventually, the information gleaned from the intercepted messages revealed the Rosenberg spy ring and exposed the atomic spy Klaus Fuchs (Hughes, 1995). Random numbers are most certainly important for the security of a nation's secrets, and more.

With the world becoming increasingly girdled by communication networks, businesses need to communicate securely with their far-flung foreign offices, and consequently industrial espionage is a growth industry. Thus inasmuch as cryptography relies upon true random numbers the importance of random numbers is growing, and, as we shall see, there are some real problems with the status quo.

Another application of random numbers comes from modern (classical) computer science. There are many kinds of computational problems that appear to be extremely difficult to solve using a classical computer executing a systematic algorithm. A case in point is the Traveling Salesman problem discussed in Chapter 3. In fact, there is no single deterministic algorithm that always finds a solution quickly for all possible problems. Recently, however, it has been discovered that one can create much more effective algorithms by adding a little randomness to the decision points within an algorithm (Traub, 1994). For example, if the program must choose between pursuing two equally promising paths, which one is it to choose? In the case of the deterministic algorithm, the choice is always the same, whereas in the case of the more haphazard "randomized algorithms," the choice will vary from run to run of the algorithm. This variable choice allows the randomized algorithm to sample effectively a much larger part of the problem in a shorter time, on average.

Typically, randomized algorithms cannot guarantee that they will solve the problem (given that a solution exists) or that they will terminate within a definite time. In contrast, deterministic algorithms can usually guarantee that they will eventually find a solution, if one exists, and that they will terminate after a definite (although typically long) time. Unfortunately, the deterministic algorithms are often extremely slow.

Although randomized algorithms give up the requirement that an algorithm *must* solve a problem, they nevertheless keep the probability of success high enough that the chances of error are small. Remarkably, such "almost correct" algorithms can be very effective in practice. Two types of algorithms that illustrate some of the tradeoffs have become popular. "Monte Carlo algorithms" are always fast and are probably correct, and "Las Vegas algorithms" are always correct and are probably fast. Depending on the application, speed may be more critical than correctness, or vice versa.

One does not usually think of the stock market as a source of random numbers. Nevertheless, it is often assumed that the stock market behaves like a "random walk," where the price from one trade to the next is not predictable. However, with systematic effects such as quarterly earnings reports, holidays, and tax deadlines, all of which affect the stock price in fairly predictable ways, we can't rely on stock prices as good generators of randomness. Furthermore, large financial institutions spend many millions of dollars to find regularities in stock- and currency-trading data. It seems doubtful that they would spend so much money on something that was completely random.

In fact, there is now a small industry of experts who peer at financial time series, charting the ups and downs of the markets in an attempt to discern some kind of pattern. In reality, the best way currently known to model the stock market is to treat it as a special kind of random process called geometric Brownian motion. Although the underlying process is fixed, many different time series are possible. Nevertheless, with a few hundred thousand sample forecasts, one can gather statistics on the expected behavior of a market and presumably make better-informed investment decisions.

Random Numbers Used in Simulations

Simulations of natural phenomena are by far the most voracious consumers of random numbers. Ever since the first "Monte Carlo" (named after the famed casinos in that city) programs of the 1950s, the need for faithful simulations has grown (Metropolis, 1953). In particular, the burgeoning field of molecular biology consumes vast quantities of random numbers in order to simulate the motion of molecules interacting with cells and other molecules.

How does randomness fit into computer stimulations? A simulation is just a mathematical model, or computer program, for some process—it is not the actual process itself. Certain simplifying assumptions about the process are made, and randomness plays a key role. To simulate the dynamics of some process, the simulation program must make choices about which outcomes will occur. For example, if two molecules collide, there is a specific probability of their bouncing away at definite angles and a specific probability of their sticking together. In the actual process there is no need for a random-number generator; things just happen. But in our computer model we use randomness as a proxy for what "just happens."

A particularly difficult computation involves determining how two kinds of molecules are likely to bind to each other. Determining the exact manner in which the two molecules twist, stretch, and vibrate as they are being bombarded by many other molecules is a horrendously complex problem. It is quite infeasible to solve it exactly. However, with the help of supercomputers and random numbers, it is becoming possible to *simulate* the motion of a molecule, in a fairly realistic chemical and biological environment, in order to determine its most likely configuration.

A detailed simulation would take into account every atomic vibration of a macromolecule. These occur at a rate of around 10^{15} *per second*. Consequently, running the simulation of a macromolecule for 100 trillionths of a second (100 picoseconds) takes about one day of supercomputer time (Hille, 1992). And this is the case even with the "simplifying" use of random numbers in critical parts of the computation!

The exact shape that a molecule adopts in a particular environment determines its medicinal and biochemical uses and, indeed, its potential toxicity. Consequently, faithful simulations of biochemical processes are providing an exciting new way to tackle drug design. Moreover, these simulations may provide new insight into the molecular-level mechanisms of diseases such as cancer and AIDS.

Random Entertainment

One neglected, though lucrative, consumer of random numbers is in the area of entertainment. We will consider two types of entertainment, casino gambling and video games. Recall that the Monte Carlo and Las Vegas algorithms were named in honor of the games played in these cities. It would seem that with all their card shuffling, dice throwing, and wheel spinning, casinos would be veritable gushers of random numbers. Is this actually true?

Sort of. Some card games can be played to the gambler's advantage, and casinos actually ban the kinds of behavior (such as card counting) that help people to win. Other games, such as roulette, seem to be truly random without any chance of someone developing a "system." However, this is not the case. Minute biases in the balance of the wheel and other idiosyncrasies can be exploited to improve the gambler's chances. For example, while at the AT&T Bell Laboratories in the 1960s, physicist Manfred Schroeder constructed a computer to predict whether a roulette ball would land on an even or an odd number on the basis of the biases in the wheel, how slowly the croupier spun the wheel, and other factors (Schroeder, 1990). Ten years later Doyne Farme, then a physics graduate student at the University of California at Santa Cruz, constructed a foot-operated computer to predict the next number in roulette (Gleick, 1987). Apparently both of these schemes worked, although neither man made much money from his ingenuity, partly because of concern that the casino management would frown on the participation of anyone who would be able to win systematically. Thus it seems that even the famous casinos in the cities after which the randomized algorithms were named are not really worthy of that honor. True random games would confer an advantage on the casinos because they would prevent clever people from guessing any biases. The challenge for the human players is to be as unbiased as the program they are pitted against—but is this possible?

In a famous experiment performed by D. W. Hagelbarger and described by Claude Shannon (Shannon, 1953), human subjects were asked to create a random sequence of pluses (+) and minuses (−). A computer program analyzed each player's past choices in an attempt to identify systematic biases among their selections. The program then predicted what symbol the player would pick next. If the player acted truly randomly, then the program should win only 50 percent of the games. In fact, when the experiment was performed, the program won 55 to 60 percent of the games. From the results of this experiment, we see that people are not very good at picking binary digits at random.

Why is this the case? As people create shortcuts to problem solving, biases and prejudices result. There are many other situations where human behavior is anything but random. For example, price bubbles in stock markets are the result of systematic delusion among people who don't want to miss out on making money. As a result, the more typically (though not completely) random behavior of the market is lost for a time.

Other biases are also seen in people's selection of lottery numbers. In fact, it is possible to write a computer program that, although it does not increase your odds of winning the lottery, can increase your odds of walking away with more of the jackpot if you *do* win by avoiding the most "popular" or "lucky" numbers that other people tend to choose!

Although this random guessing might seem like a useless exercise, in fact it could become a rigorous training method for unbiased thinking. This is not as far-fetched as it might seem. People are now shelling out money for courses in "lateral thinking" and "right-brain training," and so on. What is the goal of this training? It is precisely to reduce biases in thinking and thereby increase our ability to solve problems—or avoid getting into them in the first place! Indeed many so-called computer learning programs, programs whose performance improves with use, rely crucially on randomness for their "learning" abilities, as in the randomized algorithms discussed earlier.

Now that we have seen how important randomness is or will be to our national, physical, and mental well-being, where can we find this elusive randomness?

Does Randomness Exist in Nature?

At certain times, nature can appear to behave in quite random ways. For example, even today, despite an arsenal of supercomputers and remote-sensing satellites, meteorologists are still unable to predict the path of a hurricane accurately. Is our inability to predict such events due to ignorance, or is there some inescapable randomness in the process?

Surprisingly, classical physics places the blame squarely on ignorance and not on inherent randomness. This is because classical physics is a perfectly deterministic theory. If a classical physical system, such as a hurricane, is started off in some definite configuration of pressure, humidity, and so on, then in principle, the equations of classical physics enable us to predict exactly how it will evolve in the future. There is no randomness in classical physics. In principle, then, it ought to be possible to predict the path of a hurricane exactly. Why is it so hard to do so in practice?

In part, the problem lies in the fact that the equations governing the motion of the atmosphere are non-linear rather than linear. Small errors, or uncertainties, in specifying the initial configuration can be easily compensated for in a linear system. Any predictions of future behavior will be quite accurate. In a non-linear system, however, any errors are quickly compounded to significant uncertainties in exactly how a system will evolve. The resulting evolution may look quite random in many respects, even though it is actually evolving in accordance with deterministic equations. Such behavior is given the paradoxical name *deterministic chaos*.

A particularly simple system that exhibits deterministic chaos is the *logistic map*. The logistic map is a mathematical rule that takes an initial value and then "maps" it into a new value on the basis of a particular formula $x_{n+1} = rx_n(1 - x_n)$. The new value then becomes the "seed" for the next value in the sequence, and so on. The logistic map enables us to generate a sequence of as many numbers as we wish. For certain values of its parameters, the sequence settles down quickly into a periodic, or repeating, sequence. For example, the evolution can converge to a periodic pattern as shown in Figure 5.1. With slightly different values, however, the values in the sequence

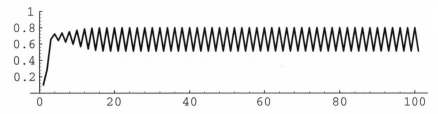

Figure 5.1 For certain values of the parameters of a logistic map, the sequence falls into a periodic pattern.

never converge to any definite pattern—they are deterministically chaotic; see Figure 5.2. Moreover, a minute change in the initial condition can quickly be amplified to a gross change in the behavior. The two curves in Figure 5.2 differ only by a change in their initial condition of one part in a billion. Yet after just 36 iterations, a significant difference in the evolution of the system is apparent.

This example shows that although there is no randomness in the equations of classical physics, it is still possible to generate evolutions that are, effectively, unpredictable because of our inability to know the exact initial condition in which some dynamical system is started. Could such processes be used as generators for sequences of numbers that would appear random? Unfortunately, in many applications that utilize random numbers, so many numbers are needed that it would be too slow and too cumbersome to use a (classical) physical process directly as a way to generate random numbers. Moreover, chaotic sequences, such as the one previously shown, harbor subtle correlations that a truly random sequence would not possess. Thus, although chaotic sequences might *appear* random, in fact they are not random at all.

What about the number pi (π)? It is widely believed that the digits of pi occur with equal frequency and would pass many tests of

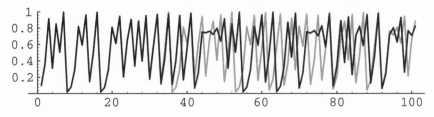

Figure 5.2 The sequence of a logistic map appears random, but the values are actually determined by a simple equation. They can be acutely sensitive to the initial condition.

randomness. Could we use these digits to make random digits for cryptographic purposes? Unfortunately, because pi is a particular number, we could in principle figure out that a sequence was coming from the digits of pi and thereby break a code based on its digits. Hence, even though the digits of pi are thought to form a random sequence in one sense, in another sense they are completely deterministic once it is known that they come from pi.

Recently, a new approach to characterizing randomness has emerged (Chaitin, 1977). It connects the idea of randomness to that of *compressibility*. The field of algorithmic compressibility attempts to characterize the smallest algorithm sufficient to generate a given sequence of numbers. Sequences that have small algorithms are deemed less random and more compressible than those that require more elaborate algorithms. By this measure, the sequences of numbers generated by the digits of pi (or indeed by the evaluation of any mathematical function) are not random, because there is a very compact rule—the algorithm itself—for describing the sequence of numbers that is generated, even though the sequence itself may be infinite. A truly random sequence is incompressible, and the smallest algorithm sufficient to generate the sequence is the sequence itself.

Gregory Chaitin of IBM has discovered a number that is the "most random" number, which he calls Ω—omega being the last letter of the Greek alphabet. Chaitin's number, Ω, is the probability that a universal Turing machine will eventually halt when fed a random bit string input (Chaitin, 1982). As there are many different universal Turing machines, there are many different Ωs. So Ω is not a single constant like π, but can be any one of several irrational numbers[1] corresponding to different universal Turing machines. In understanding the properties of Ω, however, it makes no difference which universal Turing machine is used, so people tend to refer to Ω as if it were a single number. We will follow this convention.

You can think of a universal Turing machine U as performing a mathematical operation that maps an input bit string, p, into an output bit string $U(p)$. For some bit string inputs the machine U never halts, so it is not always possible to define $U(p)$ in a meaningful way. However, we can ask a question about our chances of selecting an input that is destined to halt. In particular, we can ask

---

[1]Rational numbers are numbers of the form x/y where x and y are integers (whole numbers). The decimal expansions of rational numbers, such as $1/7, 17/35$, or, $1/(2^n)$ always repeat a pattern of digits after some point. Conversely the decimal expansions of irrational numbers, such as $\sqrt{2}, \pi$ and Ω, never fall into any regular pattern.

"What is the probability that U, fed with a random bit-string input of arbitrary length, will compute some result (we don't care which) and halt?"

To answer this question we need to specify what it means to give a computer a random input of arbitrary length. This is conceptually tricky because, without restricting the size of the input, there are infinitely many candidate bit strings to consider. To avoid this problem we imagine that the computer is not given the entire bit string initially. Instead, it requests the input program one bit at a time, and each bit is determined by the toss of a fair coin. The scheme used to select random input bit strings is designed in such a way that the probability of an input program p is $2^{-|p|}$ where $|p|$ is the number of bits in p.

To make the probabilities sum to one, the input must be "self-delimiting" in the sense that, if the computer accepts a certain input program described, for example, by the bit string 101001010010 1111101110010000010, computes some result x and halts, then no other programs beginning with the same bit string are allowed to be considered. In other words, no extension of a valid program is a valid program. If a longer program begins with the same bit sequence as a shorter one we can imagine that the shorter program is a "prefix" for the longer program. Hence, another way of describing how input programs are selected is to say that the only programs the machine is allowed to consider are "prefix-free." Summing over only prefix-free programs, the probability that the universal Turing machine computes the result x and halts is

$$\Omega = \sum_{\substack{\text{all outputs } x}} \left(\sum_{\substack{\text{all } \textit{prefix-free} \text{ programs} \\ \text{that compute } x}} 2^{-|p|} \right)$$

where the symbol $\sum_s$ means "sum over all values in the set s of . . . ." The restriction of the sum to prefix-free programs guarantees that Ω is a well-defined probability lying between zero and one. It is easy to see that Ω is greater than zero because it is defined as a sum of positive numbers. To see that Ω is also less than one, requires a little more thought. Suppose that there are N_1 1-bit programs that halt, N_2 2-bit programs that halt, N_3 3-bit programs that halt, and so on. Then the number of n-bit programs that are prefix-free, given $N_1, N_2, \ldots, N_{n-1}$ smaller programs that halt, is $2^n - \sum_{j=1}^{n-1} 2^{n-j} N_j$. Consider what Ω would be if *all* the surviving n-bit prefix-free programs halted. In this case, Ω would be given by $\sum_{i=1}^{n-1} 2^{-i} N_i + 2^{-n} \left(2^n - \sum_{j=1}^{n-1} 2^{n-j} N_j \right) = 1$. The sum is equivalent to one taken over programs of *all* possible lengths because, if all n-bit prefix-free programs halt, there are no longer prefix-free programs left to be considered. In other words the largest Ω can be, when defined as a sum over prefix-free programs, is 1. However, in reality not all prefix-free programs of a certain size halt, and so $\Omega < 1$.

So how would you compute Chaitin's number Ω? The simple answer is that you cannot compute it! Ω is uncomputable. Although Ω is a probability and must therefore lie between 0 and 1, there is no algorithm that can compute the sequence of digits. If such an algorithm existed, then the algorithm would constitute a compression of Ω and, as we have said above, Ω is an incompressible number. To see why Ω is uncomputable, imagine *trying* to compute it using the following scheme.

Begin by setting the estimate of Ω to be zero initially. Guess a random input bit string, p. Run U on this input. If U halts, add a factor of $2^{-|p|}$ to your estimate of Ω and mark all extensions of p as being "off-limits" for future consideration. Continue guessing random inputs and updating Ω. If you encounter a program that halts, which is a prefix for longer programs that have already been found to halt, strike out the contributions to Ω from the longer programs and add the contribution from the shorter one.

Unfortunately, this plausible-sounding scheme fails. The problem is that the question of whether or not a given universal Turing machine will halt on a given input is *undecidable*. In fact, this is precisely Turing's Halting Problem that we described in Chapter 3. Consequently, our candidate procedure for estimating Ω is not an effective procedure at all! We can only tell that our Turing machine will halt on a given input by running the machine and seeing it halt. If one of our inputs causes our Turing machine to run forever without halting, we can never know this no matter how long we run the machine. As a result we cannot know the factor (0 or $2^{-|p|}$) to add to our estimate of Ω. The problem remains even if we interleave random programs, taking one step of the first program, then one step of a second program and another of the first, then taking one step of a third program and another step of each of the preceding programs, and so on. This avoids the pitfall of becoming sidetracked on any one non-halting program, but it still does not allow us to compute Ω. Rather, it only gives a lower bound on Ω. Fundamentally, Chaitin's number is uncomputable because Turing's Halting Problem is undecidable.

Let us turn now to another fascinating aspect of Ω. Ordinarily we think of randomness as being the very antithesis of information. If we cannot compute the successive digits of Ω then its sequence of digits must also be random (in the sense of algorithmic information theory). However, those digits are far from being information-free. In fact, if, through some kind of mystical insight, we could come to know the true value of Ω, or even just the first few thousand digits in its decimal expansion, then we could use this knowledge to answer most of the outstanding questions in contemporary mathematics! How could this be done?

Recall that the problem of deciding the truth or falsity of any mathematical conjecture can be equated to the problem of deciding whether a universal Turing machine will or will not halt on an input that encodes that conjecture, together with the axioms and rules of inference of the formal system in which the conjecture is stated. The symbols used to pose the conjecture can be encoded on the tape of a Turing machine as a sequence of 0s and 1s.

Now *suppose* that we know Ω, the probability that the given universal Turing machine halts on a random input. Our conjecture is hardly a *random* input, of course, but it can be treated as though it were. Of all its attributes, the only one that concerns us is how many bits it consumes. Suppose that the conjecture bit string uses n bits.

Next we imagine computing a sequence of rational approximations to Ω; let's call them $\Omega_1, \Omega_2, \ldots, \Omega_n$. The rational number Ω_n is the probability that the universal Turing machine will halt on a random input of n bits or less. This number is therefore different from Ω in that it puts a limit on the size of the input bit strings to be used. However, there is a fairly simple relationship between Ω and Ω_n. It must be that $\Omega_n < \Omega < \Omega_n + 2^{-n}$. The left-hand side is certainly true because Ω_n is based on inputs of size n bits or less, whereas Ω is based on inputs of any size. Likewise the right-hand side is certainly true because an input of size n bits that causes the machine to halt would have contributed at least a factor of 2^{-n}. Such an input doesn't exist because all of them are taken into account in the determination of Ω_n. Hence, Ω_n imposes a fairly tight bound on Ω. How is this bound of use to us?

Well, we assume we know Ω (through mystical insight or whatever) and our job is to compute Ω_n. We do this using an interleaving of inputs like we saw above. Pick an input (of n or fewer bits). Take one step of the universal Turing machine on this input. Pick another input (of n or fewer bits) take one step of the new input and another step of the first input, pick a third input take one step of the third input and another of each of the second and first inputs, and so on. The inputs should be ordered so that we never pick the same input twice. Whenever we encounter an input, p, that halts, add a factor of $2^{-|p|}$ to the estimate of Ω_n (where $|p| \leq n$ is the number of bits in the input that halts). Continue this process until the difference between $\Omega - \Omega_n < 2^{-n}$.

At this point we have done something remarkable, although it is hard to see it! We have determined the bit string inputs of n or fewer bits that lead to a halting Turing machine. This is not big news. But because (by assumption) we know Ω we can also conclude that *none of the remaining inputs in* n *or fewer bits could possibly halt.* If they did halt they would contribute a factor of at least 2^{-n}, but this is impossible because we had continued the process until $\Omega - \Omega_n < 2^{-n}$ and we know for sure that $\Omega_n < \Omega < \Omega_n + 2^{-n}$. Hence, if our mathematical conjecture can be expressed in n bits or less, we will either know for sure that it leads to a halting Turing maching or that it will never halt. Hence the mathematical conjecture can be decided *if* we were given mystical insight as to the decimal expansion of Ω!

In fact, Ω tells us too much for we have not only decided on the veracity of a particular mathematical conjecture but in fact on the veracity of *all* mathematical conjectures expressible in n or fewer bits! So although the digits in the decimal expansion of Ω are random and uncomputable, they are not at all information free. In fact, the digits are so information dense, they have no extra, redundant, or useless information in them (Gardner, 1979).

Gregory Chaitin has described Ω as a number that is "suitable for worship by mystical cultists," because of its infinitely deep stock of mathematical knowledge. Although Ω is both uncomputable and random (in the sense of being incompressible), it is probably sitting out there in the Universe as we speak—it is, after all, just some irrational number between 0 and 1. Unfortunately, we would not be able to recognize it if we saw it.

Quantum Ω

As you might expect, there is a quantum generalization to Ω which we will call Ω_q. Ω_q was invented at Café Bräunerhof, a coffee house in Vienna by Gregory Chaitin, Anton Zeilinger, and Karl Svozil (Svozil, 1995). It corresponds to the *probability amplitude* with which a random quantum program halts on a universal quantum Turing machine. Thus, to determine the quantum *probability* of halting one would need to compute $|\Omega_q|^2$ (as a probability in quantum mechanics is given by squaring the corresponding probability amplitude).

However, the notion of "halting" in the case of a universal quantum Turing machine (QTM) is more subtle than that of a classical Turing machine. This is because the halt bit of a QTM might, in the long time limit, enter a superposition state of the form $a|\text{halted}\rangle + b|\text{unhalted}\rangle$ such that $|a|^2 + |b|^2 = 1$ and stay there while other parts of the output state describing the QTM continue to change. If the halt bit is in such a superposition state, although it does not change (and hence the QTM has "halted" in *some* sense) the QTM is neither truly halted nor truly unhalted. The question is only decided completely when someone measures the halt bit and either finds it in the $|\text{halted}\rangle$ or $|\text{unhalted}\rangle$ state.

The random quantum program fed into a universal QTM is assumed to be given as a sequence of prefix-free quantum bits. In other words, in computing Ω_q, only the quantum versions of classical prefix-free bit strings are allowed as inputs. This facilitates defining Ω_q by analogy with the classical case as a weighted sum over all prefix-free quantum programs that halt.

There is a slightly restricted version of Ω_q that reveals a remarkable analogy between quantum physics and quantum algorithmic information theory. The new number is Y_i, it is the probability amplitude with which a random prefix-free input halts in a particular output state $|s_i\rangle$. Like Ω and Ω_q, Y_i is also incompressible. It is written as a sum over all prefix-free inputs that halt in a certain state. Thus, it has the form

$$Y_i = \frac{1}{\sqrt{2^{|p_1|}}} QTM(p_1, |s_i\rangle) + \frac{1}{\sqrt{2^{|p_2|}}} QTM(p_2, |s_i\rangle) + \dots$$
$$+ \frac{1}{\sqrt{2^{|p_n|}}} QTM(p_n, |s_i\rangle)$$

where $|p_i|$ is the length of program p_i and $QTM(p_j, |s_i\rangle) = 1$ if the QTM running the prefix-free program p_j halts with output $|s_i\rangle$, and 0 otherwise.

It turns out that the question of whether a given input does or does not halt can shed new light on quantum physics. This is refreshing because, so far, most of the concepts of quantum computing have flowed from physics to computer science. With quantum Turing machines and Y_i, we have an opportunity to use computer science concepts to inform physical ones.

The idea is to map physical processes into halting problems of quantum Turing machines. Because a physical process has a finite duration, it must end, or "halt," at some definite time. Thus, a physical process is like a program that starts, runs, and finally halts. The bit strings supplied to the QTM represent a particular preparation of the initial state of a quantum physical process. If that physical process has several possible (observable) outcomes we can ascribe a certain probability to each outcome. Nature "running" the quantum process is like us running a QTM that simulates that process. For each possible initial state and final state of the process, we supply the QTM with a representative bit string. If the QTM halts then we know that that particular input gives rise to a physically realizable evolution. Table 5.1 shows the analogy between quantum physics theory and quantum algorithmic information theory.

Notice the similarity between Y_i (the probability amplitude/ wavefunction for a QTM halting in a certain bit string configuration) and ψ_i (the probability amplitude/wavefunction for a physical system finishing in a certain eigenstate). Just as Y_i is information

Table 5.1
Analogies between quantum physics theory and the quantum algorithmic information theory.

Analog Term	Quantum Physics Theory	Quantum Algorithmic Information Theory
System	Any quantum physical system (SYS)	Any quantum Turing machine (QTM)
Prepare	SYS in some initial quantum state	QTM in some initial prefix-free bit string state
Evolve	Into state $\lvert \Psi_{SYS} \rangle = \sum_i \psi_i \lvert e_i \rangle$	Into state $\lvert \Psi_{QTM} \rangle = \sum_i Y_i \lvert s_i \rangle$
State after observation	Find SYS in one of the eigenstates, $E = \{e_1, e_2, \ldots, e_n\}$	Find QTM in one of the bit string configurations, $S = \{s_1, s_2, \ldots, s_n\}$
Probability	Probability SYS finishes in the state $\lvert e_i \rangle$ is $\lvert \psi_i \rvert^2$	Probability QTM halts in the state $\lvert s_i \rangle$ is $\lvert Y_i \rvert^2$
Never seen	Non-realizable physical process	Non-computable result

dense with no "extra," "redundant," or "useless" information, so too is the wavefunction, ψ_i. This conclusion tallies nicely with the common saying of physicists that "the wave function contains everything we know about the system." They are quite right. But our basis for concluding that a wavefunction contains a complete, compact, and minimal encoding of the information about a physical system rests upon algorithmic information theory, not physics! It is quite striking that two radically different ways of picturing what physical evolution means predict the same thing.

Also interesting is the rather enigmatic last row in Table 5.1 marked "Never seen." Just as there are certain input bit strings to QTMs that lead to non-halting evolutions, so too must there be initial states of physical systems that lead to non-computable consequences. This does not mean such initial states cannot occur in Nature but rather that we cannot predict their outcome. In fact, it would seem that algorithmic information theory tells us that a putative "theory of everything" must necessarily contain holes. In other words, a theory of everything must admit theorems (consistent with the rest of the theory) whose veracity we can never determine by logical reasoning purely within the theory. Such implications may be verified empirically, but there is no way to deduce from whence they came by composing simpler elements of the theory. Thus a non-computable process is one that may be seen in Nature through an experiment, but not found within the theory (Pour-El and Richards, 1982).

Pseudorandomness: The Art of Faking It

Classical computers are very good at following clear, precise instructions, but they cannot be programmed to do something unpredictable, such as pick a random number. The only thing classical computers can do is evaluate functions. No matter what computational task they are asked to perform, they are always constrained to bend the problem into some function evaluation. This is all well and good if the task happens to *be* a function evaluation, but there are plenty of important computational tasks that do not require, or (worse) cannot be accomplished by, function evaluation.

Generating a random number happens to be one of the tasks that cannot be accomplished by evaluating any function. Thus a classical computer can only feign randomness, and in John von Neumann's words, "Anyone who considers arithmetical methods of producing random digits is, of course, in a state of sin."

Pseudorandom number generators all work by taking a previous value or values in the sequence and then modifying it or them in some way to generate the next number in the sequence. This bootstrapping technique is analogous to the logistic map we encountered earlier. The bootstrap is initiated with a "seed" value for the generator.

One way of generating sequences of pseudorandom integers is by using *linear congruential generators*. Linear congruential generators take as input a seed, multiplier, and modulus. The generator works according to the following program:

1. Multiply the seed by the multiplier and go to step 2.
2. Divide the result of step 1 by the modulus and go to step 3.
3. Set the random number equal to the remainder found in step 2 and go to step 4.
4. Set the seed equal to the result of step 3 and go to step 1.

The problem with linear congruential generators is that their outputs are periodic; that is, they start to repeat as soon as any number is repeated. Worse still, for many poor choices of parameters, the period can be quite small. To see this, consider the output from a particular linear congruential generator with seed equal to 1, multiplier equal to 6, and modulus equal to 11. Then the sequence generated by following the program above is 1, 6(= $(1 \times 6)/11 = 0$ + remainder 6), 3(= $(6 \times 6)/11 = 3$ + remainder 3), 7(= $(3 \times 6)/11 = 1$ + remainder 7), 9, 10, 5, 8, 4, 2, then repeats ..., which is a sequence with period 10. However, merely changing the parameters does not ensure a sequence with a large period—that is, a sequence with a large number of distinct values. In fact, for linear congruential generators, the maximal period is one less than the modulus. Keep in mind that, unlike a linear congruential generator, true random numbers *can* repeat a particular number without then having to repeat the rest of the numbers that follow.

Despite these limitations, the sequence of numbers that is output from a linear congruential generator does possess many properties that we would expect of a true random sequence.

RANDU, the random-number generator found on IBM mainframe computers in the 1960s, was based on a linear congruential generator. RANDU was thought to be adequate for many kinds of simulations that called for random numbers. However, an insidious kind of correlation was found to be lurking in the sequence of pseudorandom numbers spewed out by RANDU (Abbot, 1995). It turns out that if successive groups of three numbers from RANDU are used as a set of coordinates in a three-dimensional simulation, a remarkable pattern emerges. From most viewpoints the points appear to be randomly distributed throughout the space, but from certain orientations we can actually see that these coordinates all lie in a set of planes (Marsaglia, 1968). This effect is illustrated quite dramatically in Figure 5.3, which is hardly a testimonial to true randomness.

Shift-register generators are pseudorandom generators that alleviate the problem of linear congruential methods by using several previous values in the sequence to compute the next value in the sequence, thereby increasing the period and reducing correlations. Shift-register methods are among the best-known pseudorandom generators (Park, 1988).

Nevertheless, we have to admit by now that any purely mechanical procedure for generating random numbers is doomed to failure. However, given the speed and simplicity of computer-generated random numbers, it would be nice if we could "fix them up" to be "more random" or "random enough." One area of weakness of

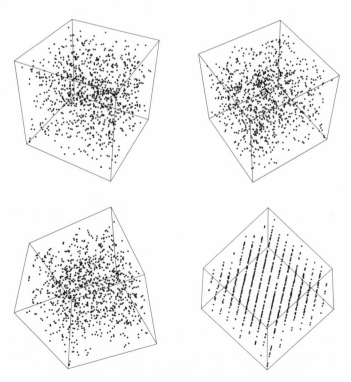

Figure 5.3 Cube of pseudorandom points seen from four different viewpoints.

pseudorandom numbers is in the way the seed value is chosen. If we know the algorithm and the seed, we can generate the entire sequence exactly, because the generator, being a function, *must* generate the same output or sequence given the same input. We can use another pseudorandom generator to generate a seed, but all we have done is to push the "seed problem" to another level.

A novel way of getting around the seed problem is to use those icons of the 1970s, lava lamps. For those not familiar with them, a lava lamp is a clear bottle filled with water and a proprietary liquid (the "lava") that forms globs that rise sluggishly to the surface of the bottle and later sink back down.

When the lamp is first turned on, a filament at the base of the lamp heats up and, after about half an hour, "melts" the lava. Because the melted lava is less dense than water, it begins to rise and, as it does so, moves in a turbulent manner. The lava continues to move viscously, subject to turbulent flow, which makes it exceedingly difficult to predict the exact motion of the lava globs. The heating at the base creates vortices in which the clear liquid and the lava undergo extremely complex thermal mixing. As the lava rises away from the heating element, it cools, oozes back down to the base, and is reheated. The cycling of globs of lava continues for as long as power is supplied to the filament.

We can use the globs of lava as the seed value for a random generator, and this has actually been done in practice (http://lavarand .sgi.com/faq.html). The random seed itself is computed by taking a digital color picture of six Lava Lite lamps. A picture containing 921,600 bytes is created, and it is manipulated and compressed into a 140-byte seed. From there the actual random number is generated by using a standard random-number generator program.

Why can't we use the output from the "LavaRand" system itself as a random number? The reason is that although the flow is quite complex, there are still regularities that make it unsuitable for a robust foolpoof generator.

The Plague of Correlations

One might think that the infamous RANDU bug is a thing of the past. Certainly, the demand for more ambitious Monte Carlo simulations catalyzed the development of supposedly "better" pseudorandom-number generators. However, *Numerical Recipes in C*, the "bible" for numerical algorithms that is widely used by scientists and engineers, reports that *"If all scientific papers, whose results are in doubt, because of bad random number generators, were removed from library shelves, there would be a gap on each shelf as big as your fist."*

This warning proved to be prophetic when, in 1992, Alan Ferrenberg and David Landau of the University of Georgia and Joanna Wong of IBM discovered a subtle bug in several supposedly high-quality random-number generators (Ferrenberg, 1992).

Ferrenberg, Landau, and Wong were interested in simulating the behavior of a three-dimensional system of interacting spins. In preparation for the full simulation, they wrote a simulator for the simpler two-dimensional case whose exact behavior had already been deduced theoretically. Thus this experiment was merely a test of the simulator against a known benchmark. Among the generators they tested were two versions of a shift-register generator.

To the researchers' surprise, the simulator gave the wrong predictions. After checking and rechecking their program, the researchers in desperation replaced the sophisticated pseudorandom-number generator with a linear congruential generator—a generator with known deficiencies.

To their amazement, the more "naive" simulator gave results that were much closer to the known answer. The conclusion was quite striking. Given current (classical) computer technology, any simulation that uses a random-number generator must be tested using different generators, regardless of how many tests the generator has already passed.

Recall the stunning illustration given in Figure 5.3 of what can happen when pseudorandom numbers are generated with a classical computer (James, 1990). The implications are important, as these next two examples illustrate.

Every year 750,000 Americans receive radiation therapy for cancer. Many of these patients die with their tumors intact. Part of the reason has to do with the difficulty in predicting how the X-rays will deposit energy in the tumor—a random process (Gibbs, 1998). Imagine receiving radiation therapy of a cancer tumor on the basis of a treatment program devised from a computer simulation of millions of X-rays that has the biases shown in Figure 5.3.

Consider also the difficulty that arises in the U.S. government's program to maintain a safe nuclear stockpile. To do this requires simulations with "unprecedented precision," all of which involve random processes in critical ways (Beardsley, 1998). Imagine the implications of assuming that nuclear weapons that relied on the pseudorandom-number generator in Figure 5.3 were "safe and reliable."

Such scenarios are unlikely, but they illustrate that random numbers are being used in important areas, so we had better get them right.

Fortunately, better and better approximations are appearing all the time (Berdnikov, 1996). But as the importance of random numbers grows, will these algorithmic generators ever be good enough? Perhaps if we looked to quantum physics, we could find a physical process that is inherently random and that could, therefore, be used as the basis for building a true random-number generator.

Randomness and Quantum Computers

Quantum-mechanical indeterminism—the inability to predict into which state a superposition will appear to collapse upon being measured—provides exactly the right solution for generating true randomness. In particular, if a physical system, representing a qubit, is placed in a state that is an equally weighted superposition of a 0 and a 1 and then measured, quantum physics says that there is exactly a 50 percent chance of obtaining either a 0 or a 1. Here is how it works.

In order to make a device that outputs a truly random bit, we simply place a simple two-state into an equally weighted superposition of 0 and 1 and then observe the superposition. The act of observation causes the superposition to collapse into either the 0 state or the 1 state with equal probability. Hence we can exploit quantum-mechanical superposition and indeterminism to simulate a perfectly fair coin toss.

By exploiting superposition and indeterminism, it is possible for a quantum computer to generate a true random number. Sequences of random numbers from quantum random-number generators will be free of the subtle correlations that bedevil classical pseudorandom-number generators.

The simple example of measuring, say, a spin state shows that there is at least one important computational task that a quantum computer can perform but a classical computer cannot.

Indeed, quantum computers may offer a way of simulating many important physical systems more directly than is possible classically. In 1982, Richard Feynman showed that no classical computer could possibly simulate a quantum–physical process without incurring an exponential slowdown (Feynman, 1982). When simulating the motion of a macromolecule in a cellular environment, a supercomputer runs about 100,000 times slower than nature. However, a general-purpose quantum computer would, in effect, be a perfect quantum simulator (Deutsch, 1985). Quantum computers might therefore be useful not only for computation but also as a way of testing various predictions of quantum physics (Wiesner, 1996).

However, we should not be too cavalier. To generate our random qubit, we still need to perform the prepare, evolve, and measure steps. Each one of these steps is subject to errors, some of which might be biased in a direction that makes our random qubits not quite so random. In fact, in the quantum world, we can never be sure of anything *exactly*. For example, we can never be sure that the superposition is exactly 50:50, nor can be we sure that our measuring devices are working perfectly properly. The bottom line is that we can never be sure we are dealing with "pure randomness" or a reasonable facsimile induced by coarse measurements, but we can at least eliminate or reduce all the known sources of problems.

The reason we have devoted an entire chapter to randomness is that it is the vital concept that distinguishes the classical universe from the quantum universe. Quantum computers can compute the necessary random numbers exactly, whereas classical computers can only approximate them. This distinction can be put to good use in generating faithful simulations of classical random processes (Zak and Williams, 1999) and in computing integrals using a quantum version of the Monte Carlo technique (Abrams and Williams, 1999). An important difference, however, is that whereas quantum computers can generate true random numbers, we can never inspect the intermediate steps they take in arriving at their answers. On the other hand, classical computers can only feign randomness using pseudo-random number generators, but we can inspect their intermediate steps. So, in terms of simulating random processes, quantum computers can give correct but unprovable results, whereas classical computers can give approximately correct but provable results. Randomness, whatever it is, lies at the heart of quantum computation.

In the next chapter we investigate yet another feat that a quantum computer can perform but a classical computer cannot: A quantum computer can give us the ability to communicate in a manner that is almost guaranteed to reveal the presence of any eavesdropping.

Six

—◦—

The Keys to
Quantum Secrets

In Nature's infinite book of secrecy
A little can I read.
—William Shakespeare

Modern schemes for exchanging secret messages, such as the one-time pad and public key procedures that we saw in Chapter 4, rely on the sender and receiver to possess certain "keys." Such keys are simply large numbers that have been carefully constructed so as to have special mathematical properties. If the appropriate keys are known, then any encrypted messages are easily unscrambled. But without the keys it is computationally intractable, at least with any classical computer, to crack a coded message. Consequently, the integrity of these cryptosystems relies on the keys being kept secret.

The problem, however, is that the keys can never be *guaranteed* to be secure. One-time pads are vulnerable to attack because before secure messages are exchanged, the sender and legitimate recipient of a message must exchange key pads (booklets of numbers) by some physical means and store them in a secure location. If the keys *could* be guaranteed to be secret, and if they were truly random numbers that were never reused, then the one-time pad would be an utterly secure cryptosystem. However, an adversary could potentially intercept and duplicate the keys at the moment they were being exchanged or afterwards during the time the key pads were in storage.

Worse still, in public key cryptosystems, the person who wishes to receive messages must broadcast a "public key" that contains a number that, if factored, would reveal the "private key," too. After the private key is known, any messages encrypted by using the matching public key are compromised. As we saw in Chapter 4, a quantum computer appears to be able to perform exactly this factoring step very efficiently using Shor's algorithm. Thus, the security of public key cryptosystems rests today on the presumption that it will be technologically difficult for anyone to build a quantum computer that can factor large composite integers—a risky assumption indeed, given the pace of technological progress.

There is a need, therefore, for a cryptographic scheme that is invulnerable to attack by an adversary who might possess a quantum computer. Such a scheme is referred to as *quantum cryptography*: the art of making coded messages by exploiting properties of quantum information—bits, quantum bits, and entangled quantum bits—that are prohibitively time-consuming to break even if a potential adversary has a powerful quantum computer.

In 1984, Charles Bennett and Gilles Brassard devised such a quantum cryptographic protocol that itself rests, fundamentally, on two quantum phenomena. In particular, this cryptographic scheme exploits the impossibility of cloning quantum information and the impossibility of measuring certain pairs of observables simultaneously (Bennett, 1992). The cloning question is related to the measurement issue, so we will focus on measurement first.

Some Underlying Concepts

The Heisenberg Uncertainty Principle

In 1927, the German physicist Werner Heisenberg discovered a fundamental limit on the accuracies with which certain pairs of observables can be measured simultaneously. That is, knowing the value of one observable more accurately necessarily makes the value of another observable more uncertain.

Many specific instances of the Heisenberg Uncertainty Principle are known. The most famous concerns trying to measure a particle's position and its momentum simultaneously (see Figure 6.1). In order to determine the particle's position accurately, it is necessary to use light with a very short wavelength. This is because the ability to provide position information is comparable to the wavelength of the object providing the position information. The problem is that short-wavelength light imparts a large momentum kick to the electron when it scatters off it to provide the position information. Conversely, if we want an accurate momentum measurement, then we have to use light with a very long wavelength, which gives us very poor position information. Similar tradeoffs arise when we measure any pair of conjugate variables.

In quantum cryptography, two people, conventionally called Alice and Bob, who wish to exchange secret messages, use the impossibility of making disturbance-free measurements on pairs of conjugate observables related to the polarization orientations of photons to establish a jointly known cryptographic key. The laws of quantum physics guarantee that no eavesdropper could *possibly* learn the identity of this key. When the key is established, Alice and Bob can then communicate secret messages using a secure classical cryptosystem such as a one-time pad. Thus, the "quantum" part of quantum cryptography lies in the key establishment and distribution step, so quantum cryptography ought really to be called quantum key distri-

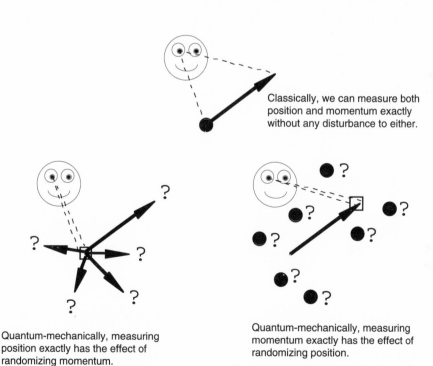

Classically, we can measure both position and momentum exactly without any disturbance to either.

Quantum-mechanically, measuring position exactly has the effect of randomizing momentum.

Quantum-mechanically, measuring momentum exactly has the effect of randomizing position.

Figure 6.1 The Heisenberg Uncertainty Principle precludes exact measurements of certain pairs of observables at the same time.

bution. Before explaining how quantum key distribution works, though, we take a brief detour into the polarization properties of photons.

Polarization

Photons are electromagnetic waves. You can think of them as consisting of an oscillating electric field and an oscillating magnetic field. The electric field and the magnetic field lie in planes perpendicular to each other and to the direction in which the photon is propagating (see Figure 6.2). One of the properties of a photon that can be used to encode a bit is the photon's polarization state. Thus, in a three-dimensional coordinate system with mutually perpendicular x-, y-, and z-axes, if a photon is propagating in the positive z-direction, the electric field and magnetic field will oscillate in the x-z plane and the y-z plane, respectively.

The photon property we are interested in is *polarization*. Photons can be linearly polarized or circularly polarized. *Linear* polarization means that as the photon propagates, the electric field stays in the same plane. In *circularly* polarized light, the electric field rotates at a certain frequency as the photon propagates. Quantum cryptography can be implemented with linearly polarized light, circularly polarized light, or a combination of the two. However, we

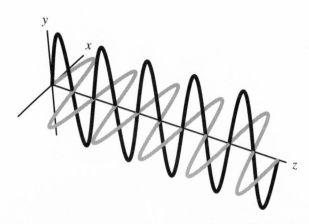

Figure 6.2 A linearly polarized photon consists of an oscillating electric field and an oscillating magnetic field that are perpendicular to each other and to the direction of propagation.

will restrict our discussion to implementations that use linearly polarized light only, because this is a little simpler to explain.

In order to encode a bit in the direction of polarization of a photon, it is necessary to place a photon in a particular polarization state. This amounts to creating a photon whose electric field is oscillating in a desired plane. One way to do this is simply to pass a stream of photons through a polarizer whose polarization axis is set at the desired angle. The only photons that emerge will be polarized in the desired orientation.

The development of modern polarizers began in 1928 with Edwin Land, founder of the Polaroid Corporation, when he was an undergraduate at Harvard College. His interest in polarization was piqued when he read about the strange properties of crystals that were formed when iodine was dropped into the urine of a dog that had been fed quinine. The crystals turned out to be made of the material that Land later used to make large-scale polarizers (Hecht, 1974).

According to quantum theory, one of two things can happen to a single photon passing through a polarizer: Either it will emerge with its electric field oscillating in the desired plane or it will not emerge at all. In the latter case, the photon is absorbed by the polarizer and its energy is re-emitted later in the form of heat.

If the axis of the polarizer makes an angle of θ with the plane of the electric field of the photon fed into the polarizer, there is a probability of $\cos^2 \theta$ that the photon will emerge with its polarization set at the desired angle and a probability of $1 - \cos^2 \theta$ that it will never emerge at all, but rather be absorbed by the polarizer.

Two polarizers with their polarization axes set at 90° from each other will not pass any light. This is consistent with quantum theory, which gives the probability of a photon emerging from the first polarizer and then passing through the second polarizer as $\cos^2 90° = 0$; in other words, because $\cos 90° = 0$, the photon is certainly absorbed.

Polarized Photons from Random Bits

A slightly more sophisticated method for placing a photon in a definite polarization state uses a device known as a Pockels cell. The Pockels cell was invented in 1893 by German physicist Friedrich Pockels, and it basically acts as a polarization-dependent switch.

By using a Pockels cell, it is possible to create a photon with its electric field oscillating in any desired plane. We can (arbitrarily) call *rectilinear* those polarized photons whose electric fields oscillate in a plane at either 0° or 90° to some reference line, and *diagonal* those whose electric fields oscillate in a plane at 45° or 135°. Furthermore, we can stipulate that photons polarized at angles of 0° and 45° represent the binary value 0, and that those polarized at angles of 90° and 135° represent the binary value 1. When this correspondence has been made, a sequence of bits can be used to control the bias in a Pockels cell and hence determine the polarization orientations from the stream of photons emerging from the cell. This allows a sequence of bits to be converted into a sequence of polarized photons. These may then be fed into some communication channel, such as an optical fiber, or perhaps even transmitted through free space.

Measuring the Polarization of a Photon

In order to recover the bits encoded in the polarization orientation of a stream of photons, it is necessary for the recipient to measure the polarizations. Fortunately, nature has provided us with a material beautifully suited for just this purpose.

A calcite, or calcium carbonate ($CaCO_3$), crystal can act as a polarization-dependent switch because it has the property of birefringence. This means that the electrons in the crystal are not bound with equal strength in each direction. Consequently, a photon passing through the calcite will feel an electromagnetic force that depends on the orientation of its electric field relative to the polarization axis in the calcite. For example, suppose the calcite's polarization axis is aligned so that vertically polarized photons pass straight through it. A photon with a horizontal polarization will also pass through the crystal, but it will emerge from the crystal shifted from its original trajectory, as shown in Figure 6.3.

This enables us to use a calcite crystal to determine whether a given photon has a vertical or a horizontal polarization. But what happens when diagonally polarized light passes through a vertically oriented calcite? The Heisenberg Uncertainty Principle says that the calcite provides no information about the original polarization, so some of the photons will not be shifted and the rest will be shifted, depending on the angle between the axis of the photon's electric field axis and the calcite's axis.

Thus we can use the location at which a photon emerges from a calcite crystal as a way of signaling (measuring) that photon's polarization.

Uncertainty Principle for Polarized Photons

To read the bit sequence "written" in a stream of polarized photons, we would have to measure (observe) the direction of polarization of each of the photons. To do this, we could use the same calcite crystal for each photon but possibly at different orientations. If we picked the "correct orientation" (e.g., if we chose to use the 0/90 orientation for a rectilinearly polarized photon or the 45/135 orientation for a diagonally polarized photon), we would get the correct answer. However, if we picked the wrong orientation (e.g., if we used the 0/90 orientation for a diagonally polarized photon or the 45/135 orientation for a rectilinearly polarized photon), then we would have only a 50:50 chance of getting the correct answer.

Now the question is whether we can measure *both* the rectilinear and the diagonal polarizations simultaneously. Unfortunately, any attempt to measure the rectilinear polarization orientation (by setting the calcite crystal in the 0°/90° orientation) *necessarily* perturbs (in fact, randomizes) the diagonal polarization orientation (the 45°/135° orientation), and vice versa. There is absolutely no hope of getting around this restriction. It is a direct consequence of the Heisenberg Uncertainty Principle, because the measurement operators for the rectilinear and diagonal polarization orientations do not commute. If an eavesdropper tries to measure the diagonal polarization when in fact the photon is in a rectilinear polarization, the eavesdropping attempt will surely randomize the rectilinear polarization state of the photon. This is the crucial physical principle that can be exploited to thwart eavesdropping during the establishment of a secret cryptographic key.

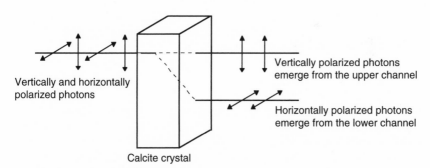

Figure 6.3 How a birefringent crystal can be used to separate photons on the basis of their polarization.

Quantum Cryptography with Polarized Photons

We now have all the ingredients needed to design a protocol for establishing a secret shared key—that is, a sequence of bits known only to the two parties who wish to communicate (conventionally called Alice and Bob).

Alice chooses a random sequence of bits out of which she and Bob will construct a key. Initially, neither Alice nor Bob has a particular key in mind. The key they will use will emerge out of the communication protocol that they will follow. Consequently, the exact sequence of bits that Alice sends to Bob is not important. All that matters is that they, and only they, come to learn the identity of a common subset of the bits. These common but private bits are used as the key.

Quantum Key Distribution in the Absence of Eavesdropping

Let us begin by considering the case in which there is no eavesdropping.

Alice and Bob need to select the probability with which they want to be able to detect eavesdropping and the number of bits they want to use in their key. These parameters determine how many photons they must exchange in order to get a key of the size they require. Suppose they would like to have a 75% chance of detecting any eavesdropping and would like to create a secure key based on 4 bits. These numbers are unrealistically low for a real cryptographic key but are chosen for the purpose of illustration.

Figure 6.4 is a diagram of the sequence of steps Alice takes to encode bits as polarized photons.

Alice chooses some set of bits (the first row in Figure 6.4). Then, for each bit, she chooses to encode it in either the rectilinear polarization ($+$) or the diagonal polarization ($\times$) of a photon (second row). This choice of polarization orientation must be made randomly. Alice then sends the photons she has created to Bob over an open communication channel (third row).

Figure 6.4   Alice encodes bits of polarized photons.

\	—	\	\	\	/	/	—	/	\	‌|	‌|	‌|	‌|	‌|	\	/	‌|	‌|	/	—	‌|	/	/	‌|	‌|	/	\	‌|	\	\
+	+	×	+	×	×	+	×	+	×	+	+	×	×	+	×	×	+	×	+	×	+	+	+	×	+	×	+	×	+
0	1	1	1	1	0	0	0	1	1	0	0	0	1	0	0	0	0	0	0	1	1	1	1	0	1	1	1	0	1	0

Figure 6.5   Bob decodes polarized photons as bits.

Next consider the actions Bob takes upon receiving the photons; these actions are shown in Figure 6.5.

Upon receipt of the photons (the first row of Figure 6.5), Bob chooses an orientation for his calcite (second row) with which he measures the direction of polarization of the incoming photons. Hence Bob reconstructs a set of bits (third row).

Now Alice and Bob enter into a public (insecure) communication in which Alice divulges to Bob the polarizer orientation of a subset of the bits. Likewise, Bob divulges to Alice the calcite orientations he used to decode the same subset of bits, as shown in Figure 6.6. For those cases in which they used the same orientations of polarizer and calcite crystal, respectively, Alice tells Bob what bits he ought to have measured. Assuming that the encoding, decoding, and transmission steps are error-free, and provided that there is no eavesdropping, Bob's test bits ought to agree with Alice's test bits 100%.

The more bits that are tested, the more likely it is that any eavesdropper will be detected. In fact, for each bit tested by Alice and Bob, the probability of that test revealing the presence of an eavesdropper (given that an eavesdropper is indeed present) is 1/4, that is, one chance in four. Thus, if N bits are tested, the probability of detecting an eavesdropper (given that one is present) is $1 - (3/4)^N$. Figure 6.7 shows how this probability of detecting eavesdropping grows with the number of bits tested. As you can see, the probability of detecting an eavesdropper approaches 1 asymptotically as the number of bits tested tends to infinity. Thus we can make the probability of detecting eavesdropping as close to certainty as we please simply by testing more bits.

1	1		0			1		0			0		0		1					1	0		
+	×		×			×		+			×		+		+					×	+		
+	×		×			×		+			×		+		+					×	+		
1	1		0			1		0			0		0		1					1	0		

Figure 6.6   Alice and Bob compare a subset of their bits to test for the presence of eavesdropping.

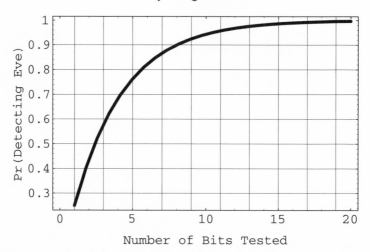

Figure 6.7 Probability of detecting eavesdropping as a function of the number of bits tested by Alice and Bob. For a bit to be tested, Alice and Bob must have used the same polarizer orientation to encode that bit and decode that bit, respectively.

After Alice and Bob have decided that the channel is secure, Alice tells Bob what polarization orientations she used for each of her bits, but not what those bits were (the first row of Figure 6.8). Bob then compares his calcite orientations with Alice's polarizer orientations (second row) and also records his own answers (third row). Bob next categorizes each bit in terms of whether he used the same orientation as Alice (fourth row). Then he looks at the bits he read for the matching cases (fifth row).

This sequence of actions allows Bob to deduce a set of bits known only to Alice and himself. To see this, compare the top line of Figure 6.4 with the bottom line of Figure 6.8. You will find that Alice and Bob agree on the bits for those cases in which they used the same orientations for their polarizer and calcite crystal, respectively.

When Alice and Bob deduce a common sequence of bits, they can use this sequence as the basis for a key in a probably secure classical cryptosystem such as a one-time pad.

×			×	×		×	+	×		+		+	+	×		+		×		+	×	×	+	+	×			×	×
+			+	×		+	×	+		+		×	×	+		×		×		×	+	+	+	×	+			×	+
0			1	1		0	0	1		0		0	1	0		0		0		1	1	1	0	1	1			1	0
☹			☹	☺		☹	☹	☺		☺		☹	☹	☺		☹		☺		☹	☹	☺	☺	☹	☹			☺	☹
				1						0								0					0					1	

Figure 6.8 Key exchange step.

Quantum Key Distribution in the Presence of Eavesdropping

Now consider what would have happened if instead there *had* been an eavesdropper, "Eve," present. Now although *we* know that eavesdropping is taking place, Alice and Bob do not, so the first step proceeds as before, with Alice encoding her bits in polarized photons, as in Figure 6.9.

This time, however, there is an eavesdropper, Eve, who is intercepting Alice's photons and making her own measurements of their polarizations in an effort to see what bits Alice is sending to Bob. Eve goes through the operations that Bob would have performed: She intercepts the photons (first row), picks calcite orientations (second row), and decodes the polarized photons as bits (Figure 6.10).

In an effort to cover her tracks, Eve then retransmits the photons she measured to Bob. Eve is free to do a complete recoding of her measured bits into photons polarized in whatever orientation she chooses, but the simplest situation has Eve using the same sequence of orientations that she used during her decoding step.

At this moment Bob is unaware of Eve's presence, so he proceeds to decode the photons that he thinks are coming from Alice but that are actually coming from Eve (see Figure 6.11). Bob intercepts the photons (first row), picks calcite orientations (second row), and decodes the photons as a sequence of bits.

1	1	1	1	1	0	0	1	0	1	0	0	0	0	1	0	0	0	0	1	0	0	0	0	0	1	0	1	1
×	+	×	×	×	×	×	+	×	×	+	+	+	+	×	×	+	+	×	+	+	×	×	+	+	×	×	+	×	×
＼	─	＼	＼	＼	／	／	─	／	＼	│	│	│	│	│	＼	／	│	│	│	／	─	│	／	／	│	│	│	／	＼	│	＼	＼

Figure 6.9 Alice encodes her bits as polarized photons.

＼	─	＼	＼	＼	／	／	─	／	＼	│	│	│	│	│	＼	／	│	│	│	／	─	│	／	／	│	│	│	／	＼	│	＼	＼
+	+	×	+	+	×	+	+	×	×	+	+	+	×	+	+	×	+	×	+	×	×	×	+	×	×	+	×	×	+	×	+
0	1	1	1	1	0	0	1	0	0	0	0	0	1	0	0	0	0	0	1	1	1	0	1	1	0	0	1	0	1	0

Figure 6.10 Eve intercepts the photons Alice sent to Bob and tries to decode them. Eve then sends the photons she decoded on to Bob, using whatever orientations she had picked to decode the photons.

│	─	＼	＼	─	＼	│	│	＼	／	│	│	│	＼	│	│	／	│	／	─	＼	＼	／	─	＼	／	│	＼	│	＼	│
+	+	×	+	×	×	+	×	+	×	+	+	×	×	+	×	×	+	×	+	×	+	+	+	×	+	×	+	×	+
0	1	1	1	1	1	0	1	1	1	0	0	1	0	0	0	0	0	1	1	1	1	1	0	0	0	1	0	1	0

Figure 6.11 Bob, unaware of Eve's presence, decodes the photons.

	1		1	0				1	0	0				0				1				0				0	
	×		×	×				×	+	+				×				+				+				+	
	×		×	×				×	+	+				×				+				+				+	
	1		1	1				1	0	0				0				1				0				0	

Figure 6.12 Alice and Bob detect the presence of Eve.

Now Alice and Bob compare the orientations of their polarizer and calcite crystal together with their intended or measured bit values for a subset of the bits Alice sent. On those cases where they agree on polarizer orientation, they should also agree on the bit sent and received. In Figure 6.12, an error in the third bit tested reveals the presence of Eve, the eavesdropper.

Consequently, Alice and Bob decide to abort their communications.

Working Prototypes

It is rare for exclamation marks to appear in the titles of scientific papers, yet in 1989 Charles Bennett and four colleagues were so excited about their achievement of building a prototype quantum cryptography machine that they wrote a paper (Bennett, 1989) entitled "The Dawn of a New Era for Quantum Cryptography: The Experimental Prototype Is Working!" David Deutsch has described this as the first device whose capabilities exceed those of a Turing machine (Deutsch, 1989).

The original prototype was built at IBM in 1989. Figure 6.13 shows a schematic of it (Bennett, 1992a). The original machine could send a secure key over a distance of only 30 centimeters. However, technological progress has been so rapid that it is now possible to transmit qubits, securely, a distance of over 30 kilometers (Marand, 1995).

The source of photons is provided by a green light-emitting diode (LED). The light from the LED passes through a pinhole, a lens, and an aperture to produce a narrow, pencil-thin beam of photons. The light then passes through a color filter to narrow the frequency spread of the green light. The light coming out of the LED is unpolarized, and the next step is to polarize the light with a linear polarizer. A pair of Pockels cells are used to generate the necessary rectilinear or diagonal polarization. The first Pockels cell can rotate a polarization by 45° when it is switched on. The second cell can rotate a polarization by 90° when it is switched on. Note that the order of the Pockels cells does not matter. Depending on which cell or cells are turned on, four possible polarization states can be generated: 0°, 45°, 90°, and 135°.

In an after-dinner speech at the 1994 Physics of Computation Conference, Gilles Brassard, who was one of the creators of the prototype, recalled whimsically that the original machine was not so secure after all: The devices used to place photons in particular polarization states made noticeably different noises depending on the type of polarization selected! So a potential eavesdropper need only listen in to the sequence of clicks made by the Pockels cells in order to learn the bit sequence Alice selected. Fortunately, such technological quirks have not impeded progress in quantum cryptography.

Since 1984 there have been several experimental demonstrations of quantum key distribution (QKD) over many kilometers of fiber optic cable (Franson, 1994; Marand, 1995; Hughes, 1995; Hughes, 1996; Muller, 1996; Hughes, 1997). This is more than enough to wire the financial district of any major city. With a QKD scheme in place, banking transactions could exploit a one-time pad cryptosystem for guaranteed-secure transactions.

However, an even more exciting possibility has emerged recently. It appears to be possible to perform quantum key distribution through the *atmosphere* with no need for special optical fibers. This is remarkable, given how "noisy" the atmosphere is to delicate quantum objects such as qubits. Free-space QKD was demonstrated over 30 centimeters by Charles Bennett and collaborators in 1991 (Bennett, 1991). More recently, James Franson's group at Johns Hopkins University has demonstrated QKD over a distance of 75 meters under bright sunlight and over 150 meters in a corridor under fluorescent lights (Jacobs, 1996). Similarly, Richard Hughes's group at Los Alamos National Laboratory has demonstrated free-

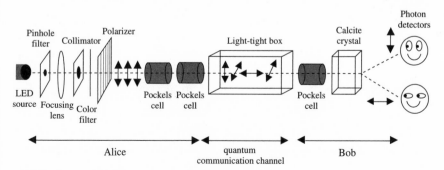

Figure 6.13 Schematic of the first quantum cryptography prototype. The actual prototype is about a meter long.

space QKD over 205 meters indoors (Buttler, 1998a) and over 905 meters outdoors at night (Buttler, 1998b).

Even more surprising is the possibility of Earth-to-space quantum key distribution. During the time it takes a satellite to cross the field of view above a point on the surface of the Earth, it ought to be possible to exchange tens of thousands of raw key bits, out of which a few thousand error-free key bits could be distilled. The distillation process is necessary to combat the effects of atmospheric turbulence and background radiation and to perform a "privacy amplification" step. If Earth-to-space QKD becomes a reality, it will allow communication with satellites that is guaranteed to be secure. Thus, experimental progress in quantum key distribution, and indeed in other forms of quantum communication, is quite staggering. There is every reason to believe that such technologies will become commercially viable early in the next century.

This chapter has described one approach to secure communications based on quantum key distribution used in conjunction with a guaranteed secure classical cryptosystem (the one-time pad). Quantum key distribution solves the problem of establishing a shared, secret cryptographic key between the parties wishing to communicate without those parties having to meet face to face. The (classical) one-time pad cryptosystem can then use such keys to establish an unbreakable classical communication channel. Together, these techniques make practical classical communications as secure as our current knowledge of physics allows.

There are now alternative schemes for quantum cryptography which exploit quantum effects other than the polarization properties of photons (Ekert, 1991; Ekert, 1992; Townsend, 1993; Townsend, 1993a). However, they all have the flavor of allowing a cryptographic key to be distilled out of a sequence of quantum bits transmitted from a sender (Alice) to a recipient (Bob). In the next chapter we shall explore an altogether different approach to moving quantum information around the Universe. This new scheme can transfer quantum information from Alice to Bob without physically moving that information through the intervening region of space that separates them. In fact, the procedure amounts to teleporting quantum information from Alice to Bob. This might sound like science fiction, but three independent experimental groups have already reported achieving teleportation in their laboratories.

So check your common-sense intuitions and prejudices at the door as we boldly go where no one has gone before.

Seven

Teleportation:
The Ultimate Ticket to Ride

*You can pull his tail in New York and
his head is meowing in Los Angeles.*
—Albert Einstein

In science fiction stories, teleportation is usually depicted as a
routine means of relocating an object by a process of dissocia-
tion, information transmission, and reconstitution. When all goes
well, the original object is scanned and disassembled at one place
only to shimmer reassuringly back into existence at another. Occa-
sional blunders corrupt the object en route or leave it suspended in
some nebulous state. Hapless bit-part actors are especially prone
to difficulties.

Such accounts of teleportation are convenient literary devices
for moving action heroes around the universe and for introducing
paradoxes of identity into story lines, but to what extent is teleportation consistent with known physical principles? In particular, does
quantum information, in the form of qubits (quantum bits) and
ebits (entangled bits) offer any new possibilities?

Until recently, no serious attention had been paid to the physical
principles on which true teleportation might be based. The presumption of most scientists, if they had any, was that teleportation
was impossible because some sort of scanning, or measurement, operation would be needed to extract a description of the quantum
state of each particle in the object. Yet the Heisenberg Uncertainty
Principle implies that such a scanning step could not be performed;

it is impossible to measure simultaneously *all* the attributes of an unknown quantum state exactly. For example, in Chapter 6 we saw that it is impossible simultaneously to measure, exactly, both the position and the momentum of a particle. Consequently, teleportation seemed doomed to failure because one could never obtain *complete* information about the original object.

The situation changed in 1993 when a team of physicists and computer scientists pooled their talents to come up with the first scientifically plausible account of how to teleport an unknown quantum state. This is considerably less ambitious than teleporting an entire human being from one place to another, but it has the advantage of being easy to analyze and compatible with known physics. The scheme allows, for example, the "spin" state of one particle to be teleported to another (possibly remote) particle, in such a way that *the actual particle does not traverse the intervening distance.* However, this operation *necessarily* scrambles the quantum state of the original particle and leaves the receiving particle in a perfect reincarnation of the original state—without ever having to explicitly determine the quantum state itself. Note that the quantum state of the original particle *must* be scrambled in the teleportation process: If it were left intact, then one could use teleportation to make a perfect copy of an unknown quantum state. But such a copy operation is expressly prohibited by the "No-Cloning Theorem." The No-Cloning Theorem says that you cannot, as a matter of principle, make a perfect copy of an arbitrary unknown quantum state.

Although the original particle and the receiving particle are distinct objects, if the two particles are the same type (e.g., if both are electrons), then it will *appear* for all intents and purposes that the original particle has been miraculously plucked from one location and deposited somewhere else, even though, in fact, no such thing has taken place.

To understand how the teleportation scheme works, we need to take a brief detour into the world or quantum entanglements and "spooky action at a distance."

Factorizable Quantum States

In science, the most profound insights often come from the simplest experiments. For example, consider a pair of anti-glare sunglasses. The anti-glare property derives from the sunglasses being made from polarizing lenses. As we saw in the previous chapter, we can think of a photon of light as electromagnetic wave whose electric field is oscillating in an arbitrarily oriented plane.

If we represent the state of a vertically polarized photon as an upward-pointing arrow in an angular box, $|\uparrow\rangle$, and the state of a horizontally polarized photon as a sideways-pointing arrow in a similar box, $|\rightarrow\rangle$, then we can represent a general superposition of these two states by the weighted combination

$$|\text{polarized photon}\rangle = \alpha|\uparrow\rangle + \beta|\rightarrow\rangle$$

in which the squares of the absolute values of the weights, $|\alpha|^2$ and $|\beta|^2$, add up to 1. These numbers give the fraction of times we would expect to see the corresponding polarization if we performed several polarization measurements on many identically prepared photons. We discussed an analogous superposition phenomenon in Chapter 2 for the case of electron spins.

Next imagine that instead of a single photon, we are now interested in predicting the outcomes of polarization tests on *pairs* of photons—we'll call them photon A and photon B. Let's create a pair of polarized photons whose polarization states are uncorrelated with each other. Thus we have four equally likely joint states: "vertical/vertical," "vertical/horizontal," "horizontal/vertical," and "horizontal/horizontal." The subscript, A or B, indicates which photon we are talking about.

$$|\text{state of pair}\rangle = \frac{1}{2}|\uparrow_A, \uparrow_B\rangle + \frac{1}{2}|\uparrow_A, \rightarrow_B\rangle + \frac{1}{2}|\rightarrow_A, \uparrow_B\rangle$$
$$+ \frac{1}{2}|\rightarrow_A, \rightarrow_B\rangle$$

The factor multiplying each pair of polarization orientations is 1/2 because we must *square* the amplitudes to get the probabilities. As there are four equally likely possibilities, there is a 1/4 chance of obtaining each one, and hence the corresponding amplitude must be 1/2.

Note that the overall state can be described as a definite state for photon A times a definite state for photon B. *Each photon has its own state independent of the other photon.* Thus, the joint state can be thought of as being factorizable as the product of a state for one photon and a state for the other, that is,

$$\frac{1}{2}\ |\uparrow_A, \uparrow_B\rangle + \frac{1}{2}\ |\uparrow_A, \rightarrow_B\rangle + \frac{1}{2}\ |\rightarrow_A, \uparrow_B\rangle + \frac{1}{2}\ |\rightarrow_A, \rightarrow_B\rangle$$

$$= \left(\frac{1}{\sqrt{2}}\ |\uparrow_A\rangle + \frac{1}{\sqrt{2}}\ |\rightarrow_A\rangle\right) \otimes \left(\frac{1}{\sqrt{2}}\ |\uparrow_B\rangle + \frac{1}{\sqrt{2}}\ |\rightarrow_B\rangle\right)$$

Each photon is therefore a superposition of a "vertically" and a "horizontally" polarized photon. This factorizability property is very important because it means that the two photons are independent of each other. Any actions we perform on one photon can have no direct effect on the other. Thus, the joint state represents a *superposition* (or blend) of all four possible pairs of measurement outcomes. There is a chance of exactly 1 in 4 of obtaining any particular pair of outcomes: vertical/vertical, vertical/horizontal, horizontal/vertical, or horizontal/horizontal. Moreover, if we measure the two photons one after another, a measurement performed on one photon does not restrict the state of the other photon in any way. To see this, suppose that we measure the polarization of photon A first and find that it is "vertically" polarized. There are now two possible circumstances consistent with this result, so the state of the pair of photons after photon A has been measured but before photon B has been measured becomes

$$|\text{state of pair}\rangle = \frac{1}{\sqrt{2}}\ |\uparrow_A, \uparrow_B\rangle + \frac{1}{\sqrt{2}}\ |\uparrow_A, \rightarrow_B\rangle$$

Note that the amplitudes changed to ensure that the probability of getting an answer overall, obtained by summing the squares of the amplitudes, is still 1.

Entanglement: Non-factorizable States

In the previous chapters, we have seen several ways of exploiting superposition and interference effects. You might be wondering what is so special about quantum mechanics if all there is to it is superposition and interference effects. We see similar effects in classical systems, too, such as when we toss two pebbles into a lake. Water waves spread out across the surface, superpose, and interfere with one another. Why all the fuss about superposition and interference effects in quantum systems?

Well, for the case of water waves, if the two waves pass through each other and spread off to different parts of a lake, you do not expect that if you dip your hand into one wave, the corresponding wave on the far side of the lake will react instantaneously in a "non-local" manner. However, this is *exactly* what happens with "quantum waves"—quantum states that are entangled. When quantum systems interact, they invariably "entangle" with one another so that they are thereafter part and parcel of the *same* single quantum system, at least until one of them is measured. Indeed, the attributes that most distinguish quantum waves from classical waves are entanglement and non-locality.

We can illustrate what is going on by using the notation we set up in the last section. For example, assume that a quantum process creates a pair of photons whose polarization states are described by the following superposition:

$$|\text{entangled pair}\rangle = \frac{1}{\sqrt{2}} \, |\uparrow_A, \uparrow_B\rangle + \frac{1}{\sqrt{2}} \, |\rightarrow_A, \rightarrow_B\rangle$$

This superposition is very special because the two polarization states are strongly *correlated* with each other. This means that as soon as we measure the polarization of one photon in the pair, we also know the result we would obtain if we subsequently measured the polarization of the other photon in the pair.

This is not the only way two systems can entangle. For example, the states $(1/\sqrt{2}) \, |\uparrow_A, \uparrow_B\rangle + (1/\sqrt{2}) \, |\rightarrow_A, \rightarrow_B\rangle$, $(1/\sqrt{2})$ $\times \, |\uparrow_A, \rightarrow_B\rangle + (1/\sqrt{2}) \, |\rightarrow_A, \uparrow_B\rangle$ and $(1/\sqrt{2}) \, |\uparrow_A, \rightarrow_B\rangle - (1/\sqrt{2})$ $\times \, |\rightarrow_A, \uparrow_B\rangle$ are equally valid possibilities. The key distinguishing characteristic of an entangled state is that it cannot be factored as the product of definite states for its sub-systems. This means that its sub-systems do not have a well-defined state until they are measured.

Given the entangled state $(1/\sqrt{2}) \, |\uparrow_A, \uparrow_B\rangle + (1/\sqrt{2}) \, |\rightarrow_A, \rightarrow_B\rangle$ if we measure the polarization of photon A and find it to be "vertical," then we know for sure that the polarization of photon B will also be "vertical." Moreover, certain pairs of measurement outcomes (namely vertical/horizontal and horizontal/vertical) are excluded from the outset.

It is important to emphasize that quantum theory tells us that the polarization states of the photons are correlated but not predetermined. That is, quantum theory says that neither photon has a definite value for its polarization until its polarization state is actually measured. Then, when the polarization state of one photon in the pair is measured, the enduring correlation between the polarization states ensures that the polarization state of the other photon becomes definite *at that instant.* There is no explanation given of how the outcome of the measurement on one photon is communicated to its twin. Surprisingly, quantum theory predicts that such an interaction is instantaneous, unmediated, and unaffected by the nature of the intervening medium. It appears to be an effect that requires no material signal to propagate from one photon to the other.

Spooky Action at a Distance

The idea that such an instantaneous, unmediated interaction existed anywhere in nature would not have sat well with Isaac Newton, one of the inventors of the calculus and of celestial mechanics. Newton once said,

> *That one body may act upon another at a distance through a vac-uum without the mediation of anything else . . . is to me so great an absurdity, that I believe no man, who has in philosophical mat-ters a competent faculty for thinking, can ever fall into.*

Nor did the prediction of quantum theory sit well with Albert Einstein, Boris Podolsky, and Nathan Rosen, who tried to use the seeming absurdity of the prediction to prove that quantum mechanics gave an incomplete description of physical reality.

Einstein, Podolsky, and Rosen believed that there were "hidden variables" that determined the polarization states of the pairs of photons from the outset. In other words, they thought that some deeper underling reality was actually fixing the outcomes of the experiments from the beginning. It was our ignorance of these hidden variables that made it appear that the states became definite upon being measured, rather than the existence of any "patently absurd" action-at-a-distance influence.

Because quantum teleportation relies crucially on such action-at-a-distance effects, it is important to digress briefly in order to convince you that the effect is real; that Einstein was wrong; and that reality does, in fact, allow instantaneous, unmediated, arbitrarily separated, "non-local" interactions.

First we had better clarify the difference between local and non-local interactions between particles. A *local interaction* is one that involves direct contact or employs an intermediary that is in direct contact. The forces with which we are familiar in everyday life, such as friction and gravity, are local interactions. With friction, the physical contact between two bodies is really mediated by an electromagnetic field, which in turn comes about by the action of an intermediary, the carrier of the electromagnetic force, called the photon. Photons travel at the speed of light, which, though fast, is still finite. Consequently, electromagnetic influences cannot propagate faster than the speed of light in a vacuum. Moreover, electromagnetic forces tend to weaken the farther we go from the source.

Locality does not necessarily imply "nearby," however. Gravity, for example is a force that exerts its influence over astronomically large distances. Nevertheless, gravity is still regarded as a local interaction because it is thought to be mediated by particles, called gravitons, which travel between gravitating objects. It too drops off in strength as the distance between the gravitating objects increases, and it too cannot travel faster than the speed of light.

An important corollary of local interactions is the following: If two events occur in regions of spacetime such that no signal, not even one traveling at the speed of light, could ever reach one region from the other, then these two events ought to be completely independent of each other. Why? Because if no signal could ever travel from one region to the other, how could what happens in one region ever be communicated to the other? In fact, special relativity has a name for two such regions; they are said to be "spacelike separated."

In short, local interactions can be characterized by three criteria: They are mediated by another entity, such as a particle or field; they propagate no faster than the speed of light, and their strength drops off with distance. Thus the assumption of "locality" allows us to infer that events in spacelike separated regions ought to be independent of one another.

Scientists have shown that all the known forces in the universe—the electromagnetic, the gravitational, the strong nuclear,[1] and the weak nuclear forces—are *local,* in this sense. One might think, therefore, that that is an end to it, and that reality must be local. After all, if *all* the known forces are local, what is left to be nonlocal?

What is left is the "collapse of the state vector." State vectors, as we noted earlier, provide the mathematical description of quantum

---

[1]Actually, the strong force is weak at *very* small distances and becomes much stronger until about the size of a nucleon (10^{-13}cm). For distances of separation greater than that of a nucleon, the strong force exerts no influence. The strong force is carried by the gluon.

systems. When we make measurements, the state vectors collapse into eigenstates. At least this is the account of measurement according to one interpretation of quantum theory. Now the intriguing point is that there is nothing in quantum theory that explains, mediates, or determines the exact mechanism of the collapse. In particular, the collapse of a state vector involves no *forces* of any kind. This lack of reliance on a force provides quantum theory with an "out"—a devious way to evade the tyranny of locality.

How exactly would a non-local influence be defined? We can just negate each criterion for a local interaction and say that a *non-local interaction* is an interaction that is *not* mediated by anything, is *not* limited to acting at the speed of light, and does *not* drop off in strength with distance. Thus non-local interactions would appear to be magic—like Einstein's poor spacelike separated cat whose pain in New York is felt immediately in Los Angeles.

Many scientists have an instinctive distaste for non-local interactions. Certainly, they would seem to be in direct conflict with Einstein's Theory of Special Relativity, which says that nothing can travel faster than the speed of light. Indeed, it was the discrepancy between the predictions of relativity and of quantum theory concerning the correlations between events in spacelike separated regions that led Einstein, Podolsky, and Rosen to accidentally discover the EPR effect whereby one part of an entangled quantum system appears to influence another instantaneously. To Einstein, Podolsky, and Rosen such non-local influences seemed implausible. They believed that the correlations in measurement outcomes when experiments measured both members of greatly separated entangled particles were more plausibly explained by hypothesizing that the pairs of entangled particles were not really entangled at all but rather had fixed values of all their measurable attributes from the outset.

Bell's Inequality

Now here comes the twist. It could be argued that it is simply a matter of philosophical *taste* whether we believe the quantum account or the hidden-variable account of how the two entangled photons come to have correlated polarization states upon being measured. But what if there were some experimentally testable difference between the predictions of the two theories? Then perhaps a physical test could resolve a philosophical question.

In the 1960s, John Bell, an Irish physicist on leave from CERN (the European Center for Nuclear Research), showed that there *is* an empirically testable difference between the predictions of any local-hidden variable theory and the predictions of quantum mechanics. The test relies on the statistics obtained when data are collected on the outcomes of pairs of polarization measurements on spacelike separated entangled particles when the polarization analyzers are oriented at certain angles to one another.

Just what would we see if we performed a set of pairs of polarization measurements? For clarity, let's suppose that the pair of photons exist in an entangled state such that both polarizations are guaranteed to be the same, but are otherwise indefinite until they are measured.

Let's call our experimenters Alice and Bob, and let's suppose that they agree to orient their polarizers in the same direction. Thus the angle between their polarization analyzers is 0°. What would Alice and Bob discover? Well, because the entangled particles we are dealing with are perfectly correlated, every time Alice observes a "vertical," Bob also observes a "vertical." And every time Alice observes a "horizontal," Bob also observes a "horizontal." The fraction of times that they agree on the measurement outcomes is 1— that is, always.

Now let's imagine what would happen if Bob rotated his polarization analyzer through 90°. Now what looks like "vertical" to Bob is actually seen as "horizontal" by Alice, so now when Alice and Bob perform polarization measurements on respective pairs of correlated photons, their results will be perfectly *anti*-correlated. Every time Alice sees "vertical," Bob sees "horizontal," and vice versa. The fraction of times that they agreed on the outcomes is 0— that is, never.

So far so good. Now suppose Bob rotates his polarization analyzer back toward Alice's vertical so that Bob's polarization analyzer now makes an angle of $\theta°$ to Alice's vertical. This is where things get interesting. Suppose the photon pair is oriented as "vertical" with respect to Alice's definition of what that means. To Bob, however, the photon he receives will appear to be a superposition of his "horizontal" and "vertical" orientations. As a result, the outcome of Bob's polarization measurement is not certain: Sometimes when Bob measures a photon that Alice sees as vertical, Bob will obtain "vertical" too. But at other times when Bob measures a photon that Alice sees as vertical, Bob will see "horizontal." The net effect is that the fraction of times Alice and Bob agree is now somewhere between 0 (never) and 1 (always), the exact number being dependent on the angle between Bob's and Alice's polarization analyzers.

In fact, just as Bob can rotate his polarization analyzer, Alice can rotate hers with respect to an arbitrary reference line that defines "vertical." We can then collect statistics on how often Alice and Bob agree on the outcome of their measurements for various angular settings of their polarization analyzers. This is not a difficult experiment to perform in principle; in fact, it seems rather mundane by the standards of modern physics.

A particularly interesting situation arises when Alice and Bob are so far apart that no signal, even one traveling at the speed of light, can possibly reach Bob from Alice, or vice versa, in the time it takes Alice and Bob to complete their measurements of the polarization orientations of their respective photons.

On common-sense grounds (as Einstein, Podolsky, and Rosen would see it), the fact that Alice and Bob are spacelike separated means that the outcome of Alice's measurement should not affect the outcome of Bob's measurement. On the basis of this assumption, it is possible to derive a rule that says how the pairs of measurement outcomes Alice and Bob see should be related to one another when Alice and Bob set their polarization analyzers at various pairs of orientations. The rule is expressed in the form that says such and such must be greater than or equal to something. It is called an "inequality," because it asserts that two things are supposed to be numerically unequal.

This rule follows logically from the laws of quantum mechanics concerning how to compute probabilities of various outcomes from

---

Bell's Inequality

The fraction of times that Alice observes "vertical" and Bob observes "horizontal" when Alice's polarization analyzer is at θ_1 and Bob's polarization analyzer is at θ_2 plus the fraction of times that Alice observes "vertical" and Bob observes "horizontal" when Alice's polarization analyzer is at θ_2 and Bob's polarization analyzer is at θ_3 *must be greater than or equal to* the fraction of times that Alice observes "vertical" and Bob observes "horizontal" when Alice's polarization analyzer is at θ_1 and Bob's polarization analyzer is at θ_3.

superpositions and the explicit assumption—in fact, the *only* explicit assumption—that reality is local. The other, "implicit" assumptions appear extremely plausible via alternative lines of evidence.

Here we have it, then: Bell's Inequality gives a definite prediction for what we should see when we measure the polarization states of pairs of entangled photons under the assumption that reality is local.

Locality: For Whom the Bell Tolls

Although John Bell derived his inequality in 1964, it was not until 1972 that anyone attempted to check it experimentally. Part of the delay was due to the inability to build good enough polarization detectors and to coordinate measurements so closely timed that no speed-of-light information could make it from one photon to the other within the duration of a pair of measurements. In addition, there was very little interest in "reality" research at the time. Most physicists were, and still are, card-carrying subscribers to the "shut-up-and-calculate" interpretation of quantum mechanics.

John Clauser, a young researcher at Columbia University, was different. Clauser took the reality question seriously. To him, and more recently to a growing number of other physicists, it really does matter what is going on behind the mathematical veneer of quantum mechanics. Calculational adequacy alone doesn't cut it. Most physicists became physicists precisely because they wanted to understand how the universe works. Comprehension, rather than calculation, was their overriding motivation. However, a physicist's training discourages philosophical musings in favor of prowess in calculation.

The results of Clauser's experiment (Clauser and Shimony, 1978), and an even more convincing version performed later by Alain Aspect (Aspect et al., 1982), are summarized in Figure 7.1. This figure shows the difference between the left-hand side and the right-hand side of Bell's Inequality as we vary two of the angles θ_2 and θ_3 for a fixed value of $\theta_1 = 0°$. If Bell's Inequality holds, then all points on the surface in the figure ought to lie above the plane through zero.

The fact that part of the surface periodically protrudes beneath this plane indicates that there are certain settings of the angles of the polarizers at which Bell's Inequality is violated. Thus the inequality is *wrong*. When the experiment is performed, nature does not obey Bell's Inequality. This means there must be some mistake in the reasoning by which the inequality was derived. But the only assump-

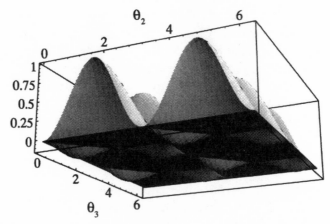

Figure 7.1 Bell's Inequality is violated for certain settings of the polarizers.

tion that was used was the assumption of locality. Hence the assumption of locality must be faulty.

Clauser's and Aspect's experiments provide strong experimental evidence that reality is non-local. In fact, rather than non-local influence being rare and esoteric events, every time particles interact with one another, their quantum states tend to entangle. Subsequently, when one member of the pair is "measured," the other member behaves as though it too had been measured and acquires a definite quantum state also. Thus, non-local influences are not the exception; they are the rule. We don't notice them in our macroscopic world, because the measurement-like interactions occur so soon after the initial entangling. But if we could scale the quantum world up to larger proportions, these exotic quantum states would be quite evident.

Can we put these strange non-local and entanglement effects to use? In the next section, we show that the answer is a resounding YES!

Quantum Teleportation

In 1993, in a paper whose author list reads like a "Who's Who" of quantum information theory, Charles Bennett, Gilles Brassard, Claude Crepeau, Richard Jozsa, Asher Peres, and William Wootters showed that it is possible to exploit entangled states and non-local influences to circumvent the limitations of the Heisenberg Uncertainty Principle and create an exact replica of an arbitrary—even unknown—quantum state, but only if the original state is destroyed in the process (Bennett, 1993). Teleportation is therefore distinct from fax transmission, which leaves the original intact and creates only an approximate replica, and from (hypothetically perfect) cloning, which would result in two identical versions of the original.

In the context of quantum teleportation, the EPR entangled pair of particles serve as two ends of a quantum communication channel: One particle is retained by the person who wishes to teleport the quantum state, the other by the person who wishes to receive it. Thus, in order to teleport a quantum state, the sender and receiver must each already possess one member of a pair of entangled particles.

Bennett and his colleagues showed that the information needed to recreate the quantum state of a simple two-state quantum system could be separated into a classical component and a quantum component, the classical part transmitted through a classical communication channel and the quantum part transmitted via (not through) a quantum communication channel, and then recombined at the intended destination to reincarnate the original state.

The basic idea is for the sender, Alice, to cause the state she wishes to teleport to Bob to interact with her member of the shared EPR pair. Then Alice makes a measurement of the resulting *joint* state of those particles and sends the result (a 2-bit classical message) to Bob over a conventional communication channel (such as a radio). Bob, knowing what operation Alice was to perform on her unknown state and her EPR particle, has precomputed a table that tells him, for each of the four possible classical 2-bit messages he can receive from Alice, what operation he is to apply to his EPR particle upon receipt of Alice's message. This allows Bob to place his member of the entangled pair of particles in a state that is an exact replica of the state that Alice wishes to teleport. The entire process is accomplished without either Alice or Bob needing to know the details of the state being teleported (see Figure 7.2).

Note that this scheme teleports the "quantum state" of an object, not the object itself. This is slightly different from the usual science fiction view of teleporting an object. Consequently, we cannot use this scheme to teleport an electron in its entirety from one place to another. Rather, we can teleport the *spin* orientation of one electron at a particular location to another electron at a different location (or indeed a different kind of particle entirely). The net effect, however, is similar: A particle in a specific state at the source loca-

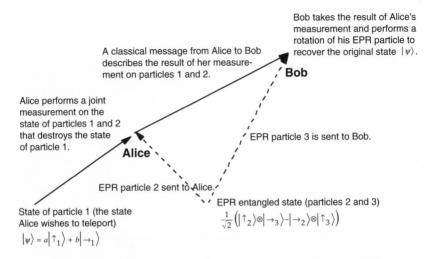

Bob takes the result of Alice's measurement and performs a rotation of his EPR particle to recover the original state $|\psi\rangle$.

Bob

A classical message from Alice to Bob describes the result of her measurement on particles 1 and 2.

Alice performs a joint measurement on the state of particles 1 and 2 that destroys the state of particle 1.

Alice

EPR particle 3 is sent to Bob.

EPR particle 2 sent to Alice.

State of particle 1 (the state Alice wishes to teleport)

$|\psi\rangle = a|\uparrow_1\rangle + b|\rightarrow_1\rangle$

EPR entangled state (particles 2 and 3)

$\frac{1}{\sqrt{2}}\left(|\uparrow_2\rangle\otimes|\rightarrow_3\rangle - |\rightarrow_2\rangle\otimes|\uparrow_3\rangle\right)$

Figure 7.2   Schematic view of quantum teleportation using EPR.

tion has its state destroyed at that location and reincarnated on another particle at the destination, without the original particle traversing the intervening distance.

Why, you might ask, would you want to teleport a quantum state? In the context of quantum computing, the answer lies in the use of quantum states to encode qubits (superpositions of conventional bits—that is, 0s and 1s). If we can teleport a quantum state between two locations, and if we use quantum states to encode qubits, then, in principle, we can teleport a qubit between two locations. Thus, teleportation might provide an alternative way to shunt quantum information around inside a quantum computer or indeed between quantum computers. This might be especially useful if some qubit needed to be kept secret. With quantum teleportation, a qubit could be passed around without ever being transmitted over an insecure (public) channel.

From a technological perspective, teleportation is much simpler than even the most rudimentary quantum computation. In fact, in 1997, two groups reported optical schemes in which they successfully teleported an unknown quantum state (Boschi, 1997) across a laboratory bench. Scaling quantum teleportation up to the level of an entire human being, however, is quite unrealistic at this point. Samuel Braunstein has estimated how much information would have to be transmitted to perform such a feat. Starting from the observation that the visible human project, sponsored by the American National Institute of Health, requires about 10 gigabytes of bits (about

16 CD-ROMs) to hold the information needed to describe the full three-dimensional structure of a human to a 1-cubic-millimeter resolution, Braunstein estimates that an entire human could be described, down to the atomic level, by using roughly 10^{32} bits (Braunstein, 1995). With current communication channel capacities, Braunstein estimates that it would take about a hundred million centuries to transmit this information down a single channel!

Moreover, the fact that quantum teleportation involves the use of a classical message limits the process to operating at the speed of light. Thus, there is no such thing as instantaneous teleportation, as some science fiction accounts suggest. However, because quantum teleportation succeeds in destroying and reincarnating quantum states, rather than the particles themselves, it does appear possible to move quantum states between different *types* of particles.

Working Prototypes

Remarkably, there are already three working prototypes of rudimentary teleportation machines. One was built by Anton Zeilinger and colleagues in Innsbruck (Bouwmeester et al., 1998), another by Francesco De Martini and collaborators in Rome (De Martini et al., 1998), and a third by Jeff Kimble's team at Caltech (Furusawa et al., 1998). There is a little rivalry between the researchers as to which machine constitutes the first *genuine* demonstration of quantum teleportation. But all three schemes are similar in their use of bench optics components such as beam splitters, parametric down converters, mirrors, and photon detectors.

A crude sketch of Zeilinger's set-up is shown in Figure 7.3.

On the left side of the figure, Alice encodes her "message" photon M toward a beam splitter, in the specific state of 45° polarization. That is, Alice intends to send the quantum state $(1/\sqrt{2}) \times (|0\rangle + |1\rangle)$ to Bob. Simultaneously, two entangled photons, A (shown as "Photon to Alice") and B (shown as "Photon to Bob"), are created and travel in opposite directions: photon A to Alice's beam splitter and photon B to Bob's beam splitter. The timings are arranged so that one of the entangled photons arrives at Alice's beam splitter at just the same instant as Alice's message photon M. Some of the time the two photons emerge from Alice's beam splitter in different directions and Alice is unable to distinguish which photon is which. As a result of this indistinguishability, Alice's message photon becomes entangled with photon A. Now neither M nor A

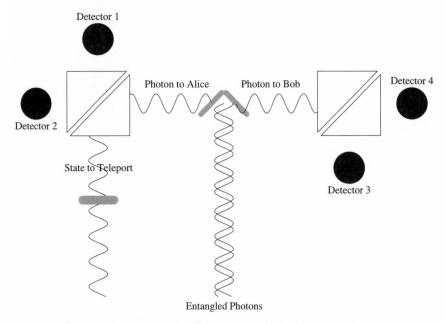

Figure 7.3   The Innsbruck quantum teleportation experiment.

has a definite polarization state, but they must be opposite because they went to different detectors when they emerged from Alice's beam splitter. However, photon *B* also has the opposite polarization state to photon *A* because photons *A* and *B* are entangled, too. Therefore, photon *B* must aquire the same polarization state as photon *M* (the message photon). Hence, teleportation is complete and Bob sees photon *B* has a polarization of 45°—the message state of a single qubit that Alice wanted to teleport to Bob.

It was quite a surprise that there was a physical basis for teleportation and an even bigger surprise that the process evolved from a concept to a working prototype in just four years. Who knows what potential this technology has for communications over the coming decades.

Eight

Swatting Quantum Bugs

I wish to God these calculations had been done by steam!
—Charles Babbage

Up to now we have blithely cavorted through idealized applications of quantum computation with little heed for what might go wrong. But things can go wrong. They go wrong in classical computers, and they go wrong in quantum computers in similar as well as novel ways. The delicate quantum superpositions and entanglements must be maintained in a bath of constantly jostling molecules, radioactive decays from the computing constituents themselves, and decays from outside sources such as comic rays. In a quantum computer the size of a piece of dust, these once happily ignored effects can throw everything completely off. These kinds of problems have led some researchers to question the feasibility of quantum computers (Landauer, 1995; Unruh, 1995). If we can't maintain the needed quantum effects at least as long as it takes to complete the computation, we're sunk. Has all the elegant theory of quantum computers been for naught?

In this chapter we will explore four different approaches to handling errors in quantum computers. The first approach is simply to do nothing and make sure that we finish our computation before errors are likely to occur. The second approach utilizes *error-correcting codes,* which involve actively correcting any errors that appear. In *fault-tolerant computing,* the quantum computing circuits are designed in such a way as to discourage errors from occurring in the first place. The final possibility is the most speculative but potentially the most robust. It is based on *topological quantum computing,* in which the physical processes themselves are tolerant to errors.

Laissez-Faire

When a new field of research is just emerging, scientists usually look for simple cases to work on, and quantum computer research is no exception. As we have seen, physicists first looked at cases where no errors occurred in the *prepare, evolve, measure* cycle. The next level of complication emerges when we consider the quantum computer to be "isolated from its environment." The phrase *isolation from the environment* refers to the "bath" we mentioned earlier; we would like to minimize such sources of error because of their debilitating effect on any quantum computation. Isolation from the environment means that any errors present at the beginning of the computation are internal to the quantum computer itself. Such errors occur, for example, when the input state is not prepared exactly as we intended. This can be particularly tricky in a quantum computer, where an arbitrary superposition can never be produced *exactly.* Nor can we even be sure that the architecture of the quantum computer is *exactly* correct, because we are talking about the positions and orientations of a large number of individual atoms. In the case of input errors, Wojciech Zurek of Los Alamos National Laboratory found that input errors do not grow with time and that architectural errors grow only with the square of the size of the error (for example, doubling the architectural errors quadruples the error in the computation) (Zurek, 1984). Both of these types of quantum computer errors grow much more slowly than in the case of the classical computer. Although these results are encouraging, they are not very realistic because of the assumption that the quantum computer is perfectly isolated from outside influences. This brings us to the next level of complication.

Over time, any real quantum system will *couple* to—i.e., become entangled with—its surrounding environment. In the process of environmental coupling, information leaks out of the qubits of a quantum memory register, and we have the beginning of an error. For example, in a spin-based quantum computer, the result of an error will be that some qubits that should have finished in the "up" position will finish in the "down" position, and vice versa. Let us now look in more detail at the kinds of errors that are most trouble-some to quantum computers.

We will use the term *external error* to describe the result of a quantum system coupling to the environment. Such external errors, as opposed to "internal" errors, arise from the surrounding thermal bath of the environment, cosmic rays, and stray gas molecules. Ex-ternal errors come in two flavors: dissipation and decoherence. In a dissipative error process, a qubit loses energy to the environment. An example is a spin value being flipped from an up to a down, or vice versa, as in the case of internal errors.

Decoherence (or, equivalently, loss of coherence) is more insidi-ous. Recall that a qubit is not just a mix of zeroness and oneness but also contains a phase factor ϕ as in $|\psi\rangle = c_0|0\rangle + c_1 e^{i\phi}|1\rangle$. Thus, rather than a simple flip of a state, decoherence causes a coupling of the qubit with the environment that tends to randomize the phase of the qubit. As a result, the interference between qubits is washed out. Thus, one of the elements necessary for quantum computation is lost. Figure 8.1 provides a graphical representation of how decoher-ence washes out a superposition. A qubit in state $|\psi\rangle = c_0|0\rangle + c_1|1\rangle$ can be described equivalently in terms of a matrix

$$|\psi\rangle\langle\psi| = \begin{pmatrix} |c_0|^2 & c_0 c_1^* \\ c_0^* c_1 & |c_1|^2 \end{pmatrix}.$$

Initially all four elements give a significant contribution, as illus-trated in the left-hand portion of Figure 8.1. However, as the qubit couples to the environment, the off-diagonal elements decay, causing the state of the qubit to evolve into

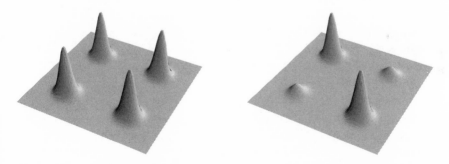

Figure 8.1 The effect of decoherence on a superposed state. Left: Initially, all states are present. Right: After coupling to the environment, some of the states (the non-classical or non-observable ones) entangle with the environment and leak out.

$$\begin{pmatrix} |c_0|^2 & e^{-\alpha t}c_0 c_1^* \\ e^{-\alpha t}c_0^* c_1 & |c_1|^2 \end{pmatrix}.$$

Thus, after enough time, the off-diagonal elements become very small, as illustrated in the right-hand portion of Figure 8.1.

The problem with decoherence is not so much that it occurs but that it occurs so quickly. The coupling to the environment is so rapid that there may not be time to perform useful quantum computations. The job for the quantum computer scientist, then, is either to devise a quantum algorithm that will finish computing before decoherence ruins quantum effects, or to lower the rate at which decoherence occurs, thereby allowing more complex quantum computations to be performed. As Table 8.1 shows, when nothing is done, coherence time is affected to an *extreme* degree by temperature and interactions with surrounding gas particles (Joos, 1985). The coherence time is how long coherence can be maintained (that is, the time before decoherence)—in other words, the time before classical behavior effects take over.

Note that in the vacuum of space, cosmic radiation does not exert a very strong decohering effect. By contrast, trying to operate a laissez-faire quantum computing system in air is nearly hopeless. At today's PC clock speeds of roughly 400 million steps per second, a quantum computer the size of a large molecule could not even compute one step before it decohered in air. That same molecule, subject only to cosmic radiation, could compute coherently for millions of times the age of the universe! Putting a dust-sized quantum computer into that same space environment, however, would reduce its coherence time to only 40 quantum computational steps.

These numbers do not assume the use of exotic materials or other specialized architectures. As shown in Table 8.2, some such

Table 8.1

Approximate coherence times, in seconds, for systems and environments of various sizes.

System Size (cm)	Cosmic Radiation	Room Temperature	Sunlight	Vacuum (10^6 particles/cm^3)	Air
Dust (10^{-3})	10^{-7}	10^{-14}	10^{-16}	10^{-18}	10^{-35}
Small dust (10^{-5})	10^{15}	10^{-3}	10^{-8}	10^{-10}	10^{-23}
Large molecule (10^{-6})	10^{25}	10^5	10^{-2}	10^{-6}	10^{-19}

Table 8.2
Maximal number of computational steps that can be performed
without losing coherence.

Quantum System	Time per Gate Operation (sec)	Coherence Time (sec)	Maximal Number of Coherent Steps
Electrons from a gold atom	10^{-14}	10^{-8}	10^6
Trapped indium atoms	10^{-14}	10^{-1}	10^{13}
Optical microcavity	10^{-14}	10^{-5}	10^9
Electron spin	10^{-7}	10^{-3}	10^4
Electron quantum dot	10^{-6}	10^{-3}	10^3
Nuclear spin	10^{-3}	10^4	10^7

materials have a much greater chance of being incorporated into actual quantum computers, as we shall see in the next chapter (DiVincenzo, 1995). As an example, consider the fourth row, which contains the data for "electron spin." In a quantum computer utilizing electron spins, the time necessary for a simple gate operation like the controlled-NOT is 10^{-7} seconds, or 100 billionths of a second. The coherence time of an electron spin—the time during which an electron is in a quantum superposition of zero and one—is about a thousandth of a second. Thus, the number of quantum computation steps that we can perform is the coherence time divided by the time needed to perform each step, which is ten thousand steps—enough to do many useful tasks even without taking any special steps to fix errors.

Note that for some materials, the characteristic time for gate operations (i.e., the time to switch between states) is very fast—in some cases a million times faster than today's PCs. But it is the product of switching speed and coherence time that determines how many coherent steps can be performed, and this is the key to whether quantum computers will ever be viable. These results are quite encouraging; it seems that a significant number of computational steps can be performed before quantum coherence is lost. Whether general-purpose quantum computers can be built from these materials will be considered in the next chapter.

The laissez-faire method of dealing with errors is one of passive acceptance of the inevitability of decoherence. If quantum computers are to be useful general-purpose computing machines, we may have to take a more active role in maintaining coherence. Our choices for proactive decoherence management are to suppress decoherence directly or to undo its effects once it has occurred. This brings us to the topic of error correction.

Error Correction

The Classical Case

Before discussing the more elaborate process of quantum error correction, let us first review how error correction is done with today's classical computers. There are no phase problems with classical bits because they have no phase to consider. Classically, we can measure the value of a bit without disrupting it and then change its value if necessary. If a bit is implemented as a voltage level of a transistor, then there are only two valid values for that voltage: a voltage corresponding to a 0 and a voltage corresponding to a 1. No other values are permitted. In reality the voltage will drift around somewhat, but the voltage difference between 0 and 1 is so large that it is clear what the actual value ought to be, even in the presence of drifts.

How can we tell if there is an error to begin with? One straightforward way of doing this is to assign several computers the same task. If the computers are executing the same deterministic algorithm, then they should produce the same answer at each step in the computation. Thus, if we poll the computers periodically and ask what their answers are at a particular stage in the computation, we can use the "majority vote" to ascertain which answer is correct. If the chance of an error per computer is less than 50 percent, then the chance that the majority vote is correct is more than 50 percent. Another twist on this redundancy scheme is to have different people write software that is supposed to implement the same algorithm. If the answers are different, then there could be a software error rather than a hardware error, but the same majority-vote technique can still be used. One constraint on the use of majority-vote schemes is that the "voters" must be independent. For example, suppose all the computers relied on the same power supply. If that power supply became erratic, then *all* the computers connected to it might produce bad results, and hence even the majority vote could be incorrect. Redundancy-based error-correction schemes are used where reliability is crucial, such as programs used aboard the Space Shuttle.

Table 8.3 shows how majority voting corrects errors in the case of three bits. The majority vote is determined in three steps. Each step involves a pairwise comparison of two bits with the other bit. The bit being tested is flipped (a 0 becomes a 1, and vice versa) if the other two bits agree with each other but not with the bit being tested. Otherwise, when the two bits used for comparison disagree with each other, the bit being tested is left alone. Three bits are the minimum to guarantee that a majority decision can be reached. That is, after the majority vote, the state of the three-bit system will be either 000 or 111.

As a concrete example, reading along the third line of Table 8.3 we have an initial state of 010. In the first pairwise comparison, we see that bits 2 (= 1) and 3 (= 0) disagree with each other, so we leave bit 1 (= 0) as it is; the overall state is still 010. In the second comparison, bits 1 and 3 are both 0, so we change the state of bit 2 to a 0 as well. The state is now 000. In the final step (which is not needed in this case); we compare bits 1 and 2. They are both 0, so bit 3 must also be a 0, but because it is already a 0 from the previous step, we don't have to change anything. The state remains 000. Thus, we have changed an initial erroneous state with one bad bit to a state where all the bits are in agreement.

In majority voting, the more computers, the better the chances of detecting any errors. Figure 8.2 shows how adding more computers increases the chances of obtaining a correct majority vote.

Table 8.3
Majority-vote error correction with three bits.

Initial State	Bit 1 Compared with Bits 2 and 3	Bit 2 Compared with Bits 1 and 3	Bit 3 Compared with Bits 1 and 2
000	000	000	000
001	001	001	000
010	010	000	000
011	111	111	111
100	000	000	000
101	101	111	111
110	110	110	111
111	111	111	111

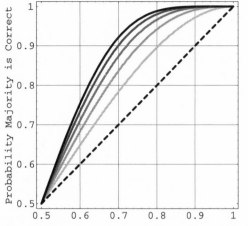

Probability Single Computation is Correct

Figure 8.2 Probability that the majority vote is correct based on the probability of a single independent computation being correct. The dashed line is the case of a single computer, the lightest curve is for three computers, and the darkest curve is for 13.

The Quantum Case

The most obvious approach to quantum error correction is recycling techniques from classical error correction to see how efficacious they are for quantum computers. But it should come as no surprise that the story is much different in the alien realm of quantum computing. First, we cannot measure the qubit without disturbing its value. Second, the amount of zeroness and oneness that a qubit may have is arbitrary and takes on a continuum of values, but when we measure it, we obtain a 0 or a 1 and lose the superposed values of zeroness and oneness. Thus, a *direct* application of majority voting is not useful, because the very act of determining the majority vote would alter that vote.

However, in certain special situations it is still possible to use majority voting. For example, suppose our computation was such that at certain predictable times, the states of the qubits were totally classical, either a $|0\rangle$ or a $|1\rangle$. Thus, if we were to measure the qubits at precisely the times at which we expected the qubit to be wholly $|0\rangle$ or $|1\rangle$ they would already be "naturally" collapsed, even though at the next step in the computation they would again be superpositions of 0 and 1. Because they are already collapsed, our measurement of them can cause no harm to the superposition, which at that moment is just one of the two states—assuming that the measurement itself is not so harsh as to disrupt the subsequent computation. Thus, only in special situations is a majority-vote scheme useful for correcting the state of a quantum computer *during* a computation.

Symmetrization

As we saw in the previous section, the use of majority-vote error correction is quite limited for quantum computers. Fortunately, there is a much cleverer way to correct errors that is simply not available to classical computers. This technique is called error correction via *symmetrization* (Barenco et al., 1996). The technique uses n quantum computers, each performing the same quantum computation. Provided no measurements are made and no errors occur, all n computers will evolve according to the same Schrödinger equation, so they will, at all times, be in the same quantum state. Moreover, as the n computers are independent of one another, their joint state will be describable as the product of the states of the n computers. Hence, if we shuffle the computers, we should see no net change to their joint quantum state. This means that the "correct" quantum state lies in some computational subspace that is symmetric with respect to an interchange of computers.

Now consider what happens if an error occurs in one of the n quantum computers. The symmetry of their joint state under a shuffle operation will be lost because the buggy computer will have a quantum state that is distinguishable from that of the other computers and so its position in the shuffle will spoil the symmetry. But if we could make a measurement that projects the joint state back into the symmetric subspace, we should nevertheless be able to undo the error, even without knowing what the error is.

This is exactly what quantum error correction by symmetrization accomplishes. Through a careful entangling of the n computers with a set of ancilla qubits representing all possible permutations of n objects, the computations can be performed over all permutations of the computers simultaneously. Then, by measuring the ancilla qubits, the joint state of all n computers can be repeatedly projected back into just the symmetric computational subspace, thereby thwarting any errors that might have arisen in one of the computers.

We can get an idea of how error correction by symmetrization works by imagining tumbling a wet sponge. We can think of the sponge as being analogous to the "computational subspace" in which the computation is supposed to take place. The water is analogous to the computation. With the sponge held stationary in our hands the water starts to drip out because of gravity. This is analogous to an error causing the computation leak out of its computational subspace. Fortunately, in the case of the sponge, we can compensate for the errors due to gravity by turning the sponge over and over in our hands. If the sponge is shaped like a cube we might lose track of which face we are looking at but this doesn't matter so long as each face points toward the ground for an equal length of time. That is, provided the sponge is symmetric and we tumble it just quickly enough to keep the water contained within it. Similarly, we can compensate for errors in the computation by continually projecting the computation back into its symmetric subspace. The ancilla qubits have the effect of performing all possible permutations on the quantum state of the quantum computers just as out hands perform all possible permutations on the orientation of the sponge.

Error-Correcting Codes

So-called error-correcting codes are used routinely in classical computers to correct for drastic errors such as bit flips. In their quantum analogs, the key is to "fight entanglement with entanglement." (Preskill, 1997a). Whereas we have seen how entanglement with the environment is a bad thing, entanglement according to our own design can be our friend. What we do is spread out the information contained in a qubit over several qubits so that damage to any one of them is not fatal to the computation. More technically, the basic idea is to encode each bit in a set of codeword bits. The codewords are chosen so as to be maximally distinguishable from each other. If an accidental bit flip occurs in a codeword, then the correct codeword (and the correct bit) can be recovered by choosing the nearest codeword.

The quantum version of error correction via coding is a little different. The problem is that we are not allowed to read the state of the buggy quantum computation directly. To do so would cause the superposition to collapse in some unpredictable way, thereby randomizing whatever remnants of correct information there were buried in the buggy state. This would make the error worse rather than better. In fact, the prohibition on reading the buggy state directly led several researchers to speculate that quantum computers would never be feasible experimentally, because they seemed to require absolute perfection in fabrication, initialization, operation, and readout—none of which are possible in practice.

In 1995, this perception changed when Peter Shor devised an algorithm that could correct an error in one logical qubit (i.e., a qubit participating in a computation) by entangling it with eight other qubits (Shor, 1995). The sole purpose of these extra qubits was to spread the information in the state to be protected among the correlations in an entangled state. This 9-qubit code was later improved upon by Raymond Laflamme, Cesar Miquel, Pablo Paz, and Wojciech Zurek of the Los Alamos National Laboratory. Their algorithm uses only five physical qubits to encode one logical qubit (Laflamme, 1996).

The key trick that Shor and these other researchers realized was that we do not need to *know* the error in order to *correct* the error. Instead, it suffices to entangle the logical qubit to be protected with several ancilla qubits, allow an error to occur among the entangled qubits, and then discover how to undo it by measuring the ancilla qubits to reveal an *error syndrome*. For each possible error syndrome there is a particular operation to be performed on the remaining (unmeasured) logical qubit that will restore it to its bug-free state (without ever needing to know what the bug-free state is).

Thus the first step in protecting a logical qubit is to entangle it with several ancilla qubits, by pushing them through an encoding circuit, such as that shown in the left-hand side of Figure 8.3.

The encoding circuit is read from left to right. On the left-hand side of the circuit is the initial state of the logical qubit, $|\psi\rangle$, and the initial states of the ancilla qubits, all $|0\rangle$. Each horizontal line indicates how a qubit "flows" through the quantum circuit. Various

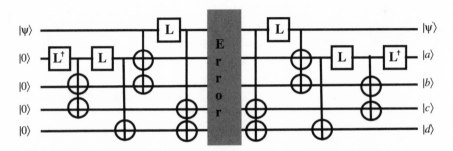

Figure 8.3 Quantum circuit for encoding and decoding a single qubit for identifying an error syndrome.

qubits are coupled, via the action of quantum logic gates (analogous to the action of NOT and AND gates in conventional Boolean circuits). The gates that are drawn as circles with crosshairs are controlled-NOT gates, and the gates that are drawn as square boxes are rotations of a single qubit. The encoding operation is the sequence of quantum gates from the left-hand side of the circuit up to, but not including, the gray region in the middle of the circuit. The quantum state that emerges after the encoding operation is an entangled state of all five qubits. Now when an error afflicts any one qubit in this set it can be undone, making the group of five entangled qubits less susceptible to error than the single logical qubit would have been if left unprotected.

To give a specific example, imagine that an error occurs in the gray region in the center of the circuit. On the right-hand side of Figure 8.3 is the decoding circuit where the error-immunizing entanglements are undone. Upon reading the state of the ancilla qubits in the output, at the extreme right of the figure, we obtain a particular error syndrome and hence the appropriate corrective action. For example, if the ancilla qubits are found to be in the states $|1\rangle|1\rangle|1\rangle|1\rangle$, then the appropriate corrective action to be applied to the final data qubit $|\psi'\rangle$ would be a rotation described by the unitary operator $\left(\begin{smallmatrix} 0 & 1 \\ -1 & 0 \end{smallmatrix}\right)$.

The quantum error-correction schemes we have described can take into account a single bit flip error, a single phase shift error, or a combination of the two. But what happens if there is more than one such error? How many errors can be tolerated and we still be guaranteed of recovering our original qubit? We can find theoretical bounds, but there is no guarantee that there is a physically realizable implementation. To get an idea of the magnitudes of the numbers involved, see Figure 8.4 (Ekert, 1996).

In the figure, the axis labeled "l" is the number of qubits used to represent data. The other axis, labeled "n," is the total number of qubits (data plus entangled ancilla qubits) needed to ensure a correct preservation of the data qubit. The vertical axis is the maximal number of errors that can be tolerated without an error. For example, reading along the axes by counting the squares, 10 qubits coded into a total of 15 qubits cannot tolerate even a single error. That is, we cannot recover the original data. However, adding one more entangling qubit means we can tolerate one error among the qubits. The squares at the lowest level, the "floor," are the cases where no errors can be tolerated, at the first "step" one error can be tolerated, at the second "step" up to two errors can be tolerated, and so on.

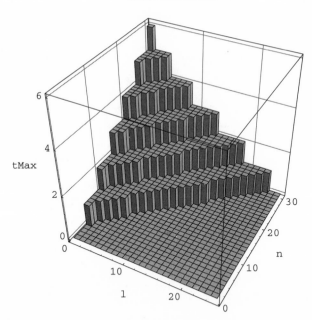

Figure 8.4 Graphical representation of the relationship between the number of data qubits ("l"), the total number of entangled qubits ("n"), and the maximal number of errors that can be tolerated ("tMax").

Concatenated Codes

The field of error correction is one of the most active areas of research in all of quantum computing. One of the most intriguing recent discoveries involves *concatenated codes*. In these codes, a coded qubit is itself further encoded within a hierarchical tree of other entangled qubits. A schematic of this structure is shown in Figure 8.5.

To be more specific, consider the case where each qubit is encoded as a block of 5 qubits. Each qubit within the block of 5 is encoded in another block of 5, and so on, until some number of hierarchical levels is reached. Each 5-qubit block can recover from one error. If we concatenate a block—that is, embed it so that we now have 5^2 qubits—then an error occurs only if two of the sub-blocks of size 5 have an error. Similarly, if we had 5^3 qubits, we could have errors in three of the sub-blocks and still compute reliably.

Remarkably, if the probability of an individual error is pushed below a critical threshold, then the quantum computation is theoretically correctable indefinitely. Hence this hierarchy of entanglements can be used to compute forever without error. If the error rate is be-

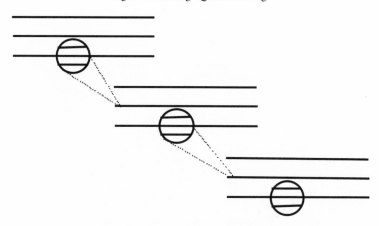

Figure 8.5   Schematic view of concatenated coding.

low the threshold, then we can simply add more levels of concatenation to reduce the overall chance for an error to an arbitrarily small level. This means that we could *in principle* quantum compute forever without an error (Preskill, 1997b). Of course, the number of qubits increases exponentially.

Fault-Tolerant Computing

In the error-correcting schemes we have discussed so far, we have implicitly assumed that the correction is performed flawlessly. However, the error correction itself is a quantum computation, so it too will be prone to errors. To utilize error correction properly, we need to make sure that it is being administered properly. *Fault-tolerant* is the term used to describe a computer that operates effectively even when its components may be faulty. John Preskill of the California Institute of Technology has identified five "laws" of fault-tolerant computing to make possible effective error correction in quantum computing (Preskill, 1997b).

1. Don't use the same ancilla qubit twice. The idea here is that if there is an error in the ancilla qubit, then this error may propagate to other parts of the network and corrupt good data qubits. An example of a good scenario and an example of a bad scenario are shown in Figure 8.6.

2. Copy the errors, not the data. We need to be careful to prepare the ancilla qubit in such a way that when we measure the an-

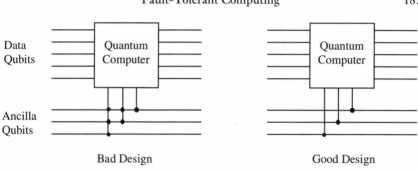

Figure 8.6 The first law of fault-tolerant computing: (left) A bad error-correcting scheme. (right) A good error-correcting scheme.

cilla in preparation for some sort of error-correcting operation, we acquire only information about any errors and nothing about the data qubits with which the ancilla is entangled. A diagram of this is shown in Figure 8.7.

3. Verify when encoding a known quantum state. The step of encoding is sensitive and prone to error. If we do not check that the encoding has been done correctly, any errors will propagate to other parts of the circuit with disastrous consequences. A diagram of this is shown in Figure 8.8.

The figure is read from left to right. In this example, three qubits equal to 0 are encoded into a block-of-three state called Qblock. In the next step, the state of Qblock is verified by first entangling it with ancilla qubits and then measuring the ancilla. This step is repeated in the next block. Assuming that both of these steps have

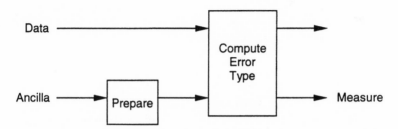

Figure 8.7 The second law of fault-tolerant computing: The ancilla must be prepared properly so that the data are not corrupted when measured. (Adapted from Preskill, 1997b.)

```
qubit1=0 ─────┐                ┌──────┐                ┌──────┐
              │       Qblock   │Verify│   Qblock       │Verify│
qubit2=0 ─────┤Encode├─────────┤Qblock├────────────────┤Qblock├──────── Qblock
              │       │        │      │                │      │
qubit3=0 ─────┘       │  Anc ──┘      └── Anc ──┐       └──┐
                                               │          │
                                             Meas        Meas
```

Figure 8.8 The third law of fault-tolerant computing: Verify the state. (Adapted from Preskill, 1997b.)

verified the state of Qblock, we can be confident that what we have at the end of the circuit is a block code of a 0.

4. Repeat operations. Just because we have verified the encoding of a state, that does not mean that it is correct. It would be just as disastrous to correct an error or non-error in the wrong way as to miss an error in the first place. By repeating measurements, we increase our confidence that the error syndrome is actually what we think it is. A diagram of this is shown in Figure 8.9.

Again the circuit is read from left to right. In this case we perform two checks of the error syndrome, using entanglement with the ancilla and subsequent measurement of the ancilla. If necessary, we also perform a recovery operation.

5. Use the correct code. This is the actual procedure that we use to correct the error syndrome. Different codes may be available, but not all are efficient and not all observe the previous four laws.

Obeying these five laws of good design will help to ensure that a quantum computer will operate reliably. Still, these techniques are somewhat cumbersome. Is there a more efficient means of ensuring correct quantum computation?

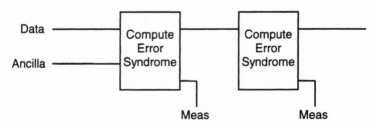

Figure 8.9 The fourth law of fault-tolerant computing: Repeat operations. (Adapted from Preskill, 1997b.)

Topological Quantum Computing

Perhaps the most fascinating method of dealing with errors in quantum computing is topological quantum computing. Rather than relying on special materials or architectures, topological quantum computing utilizes fundamentally different quantum *processes* that are more immune to the vagaries of unwanted decohering entanglements. That is, the quantum gates themselves behave in a way that is intrinsically tolerant to errors, so we don't have to rely on special circuit connections and manipulations to achieve fault tolerance.

Topological quantum computing takes advantage of the non-local characteristics of quantum interactions: An electron can "be" in more than one place at the same time because multiple paths are possible quantum mechanically. By the same token, if we encode quantum information in a more global fashion, then local disturbances will not affect the quality of the computation. Is there such a quantum interaction that may useful here?

The Aharonov–Bohm effect is one such effect with the desired properties (Preskill, 1997a). Figure 8.10 shows an electron that travels around a tube of magnetic flux. In traveling around the tube, the electron spin undergoes a phase shift. But what is interesting is that no matter what path the electron takes in its circumnavigation around the magnetic field tube, the phase shift is the same. The Aharonov–Bohm effect has an infinite range, so the electron can in principle be infinitely far away and not change its resultant phase

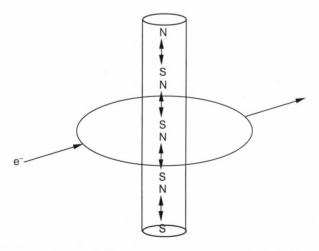

Figure 8.10   The phase shift depends only on the number of turns around the flux tube, not on the path.

shift. The only thing determining the phase shift is how many times the electron goes around the tube.

In terms of quantum computing this means that if local disturbances were to change the path of the electron, these disturbances would not affect a computation that depended only on how many times the electron went around the tube. Thus, such an electron could literally orbit Earth and still have the same effect on the computation as if it described a nanometer-sized orbit. This exceedingly robust technique could be embodied as spins arranged on a surface. Alexei Kitaev of the Landau Institute for Theoretical Physics in Moscow has shown that it is possible to construct universal quantum gates using only Aharonov–Bohm interactions (Kitaev, 1997).

The theoretical results outlined in this chapter, and a growing number besides, point to the fact that the errors that inevitably arise in delicate quantum computation may not be so disastrous as was once believed. These results encourage the hope that through a judicious choice of naturally decoherence-resistant materials, concatenated codes, error-correcting codes, and topological computing schemes, we can make quantum computers a viable technology. Perhaps if Charles Babbage were alive today, he might very well say, *"I wish to God these calculations had been done by quanta!"*

Nine

Generation-Q Computing:
Where Do You Want to Go
Tomorrow?

It means you can try to answer questions you thought
the Universe was going to have to do without.
—Bruce Knapp

It should be clear by now that quantum computers are theoretically possible. The questions are whether they can be built and, if so, how to build them. Throughout this book we have been discussing the one-atom-per-bit limit that, judging on the basis of current extrapolations, will be reached by the year 2020. If that goal is two decades away, why should we be worrying today about how to build quantum computers? The reason is that quantum computers need not wait for the one-atom-per-bit limit to be reached. Just as transistor technology achieved astounding miniaturizations from the first coin-sized transistors, we can expect the same sort of evolution from quantum computers. As we will see, true quantum switches and logic gates have already been built. These tabletop-sized devices are obviously not feasible for any practical quantum computer, but they represent the necessary first steps in the process.

These first steps in the process of building fully functional quantum computers have been taken using radically different techniques. It is still quite early in the venture to know which of these techniques is likely to "win," but we will point out the ones that at this time appear to be reasonable candidates. And as often happens in science, an entirely new technology that lies just ahead may make all the existing techniques obsolete.

We can divide today's promising quantum computing technologies into three approaches. The first approach is reminiscent of the way today's conventional computers operate. Such quantum computers would be more like quantum extrapolations of today's classical computers, computing through the use of general-purpose circuits. One of the realizations of this method goes by the name of *ion traps*. The second approach involves very simple processors with unprecedented redundancy. In this way, we need not worry so much about maintaining the delicate correctness of ephemeral qubits. The elaborate machinery needed to maintain correctness is replaced by an average result of the many processors, and this result achieves the same levels of accuracy as those qubits that require meticulous restoration. This approach utilizes nuclear spins in a technique called *nuclear magnetic resonance,* or *NMR.* The third approach builds upon decades of silicon-based semiconductor technology. It envisions using individually addressable nuclear spins as the qubits. Conditional quantum gate operations are achieved by coupling the nuclear spins indirectly via electron spins.

All three approaches are appealing because they are based on tried and true principles. Also, a large, existing infrastructure and knowledge base could be relatively quickly tuned to the new technology. Such methods therefore have the potential to yield a more rapid return on investment. The transition from classical to quantum physics in the first 30 years of the twentieth century followed this "retuning" method as it reified the dusty old classical equations into shiny new quantum ones. The downside of extrapolating proven methods is that we may become unnecessarily stuck in a suboptimal technology from which it may take us decades and many billions of dollars to recover. Nevertheless, let us begin by examining the implementation of quantum logic in hardware.

Quantum Conditional Logic

Quantum and classical computing technologies share a number of basic elements. One of the key ingredients of quantum computers is the notion of *conditional logic*—that is, a set of rules for a logic gate whose output depends not only on its input but also on the state or condition of other gates.

A group from the Institut für Theoretische Physik at the University of Stuttgart, W. G. Teich, K. Obermayer, and Günter Mahler, have designed a molecular quantum switch that we use as the basis for our discussion (Teich, 1988). Their model was built upon by Seth Lloyd, at M.I.T., who extended the model to include superpositions and thus true quantum computations (Lloyd, 1993). Both models can be extended to use a heteropolymer—that is, a molecule of different atoms that repeat their order—as the substrate for a multiprocessor system that acts as the hardware for the computer. So-called *quantum dots* or ion traps work in this architecture as well. A quantum dot is an electron confined to a small region where its energy can be quantized in a way that isolates it from other dots or ions. The "corral" that contains the electrons uses "fences" of electromagnetic fields either from other atomic electric fields or from waves of photons that add constructively to create a barrier for the ion. In the NMR realization of a quantum computer, coupled nuclear spins of atoms in molecules provide the necessary quantum computing mechanism and could be made with a heteropolymer as well. Because the basic phenomena are more or less similar for all these embodiments, we will explain these phenomena in some detail to provide a foundation in the basics of how these quantum computer components operate.

For purposes of illustration, let us consider a linear collection of atoms to be memory cells. Each atom can be in either of two states, an excited or a ground state, which constitutes the basis for a binary arithmetic. Neighboring atoms interact with one another in a way that enables a conditional logic to be supported. For example, an atom in an excited state that has both neighbors in the ground state will behave differently from an atom whose neighbors are in different excited states, either through an electrical charge coupling (that is, an attraction or repulsion) or through a coupling between spins. The collections of atoms and their possible states constitute the "hardware" for the computer. The software for the ion trap computer consists of a sequence of laser pulses of particular frequencies that induce transitions in certain atoms and thus change the values of the data they represent. In the NMR quantum computer, the laser pulses are replaced by radio frequency (RF) pulses that rotate or flip the nuclear spins. Both laser pulses and RF pulses are forms of electromagnetic energy—that is, photons.

Thus, in both the ion trap and NMR models, pulses of electromagnetic radiation provide a mechanism for initializing a qubit in a certain starting state and for changing its state during computation. However, unlike classical computers in which each bit can only ever by in the 0 or 1 state, by varying the length of the electromagnetic pulses we can place a qubit in an arbitrary superposition of 0 and 1 simultaneously. When we augment this capability with conditional logic operations between different qubits, we can then entangle sets of qubits and perform general quantum computation.

In the Stuttgart model, each atom has a three-energy-level profile, as shown in Figure 9.1. The purpose of the additional third level is explained below.

State 0 is the lowest or *ground* state, state 1 is a *metastable* excited state, and state 2 is an excited state rapidly decaying to either state 0 or state 1. *Metastable* means that the transition between state 1 and state 0 is very slow compared to the transition from state 2 to state 0, so for practical purposes we can ignore this slow transition. In a computer, the "practical purpose" is that the processing done with the atom in the metastable state takes place before the transition to the ground state 0 is likely to occur. As with all quantum phenomena, we don't know exactly when it will decay; we know only the probability that it will decay within a certain time. There is always some chance for the transition to take place, even if we have not instigated it as part of our computation.

Transitions between the atomic states are accomplished with laser pulses of the appropriate frequency. For example, the transition between state 2 and the ground state proceeds in either direction (from 2 to 0 or from 0 to 2) when a laser pulse of photons has an energy corresponding to the difference between states 0 and 2 or, equivalently, a frequency of $f1$. In the case of the transition from state 0 to state 2, the laser photon is absorbed, and in the transition from state 2 to state 0, the laser photon acts to stimulate the transition, which results in the emission of a photon of the same energy as

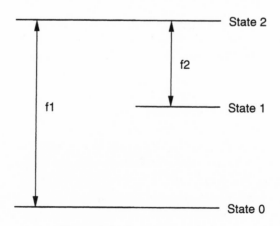

Figure 9.1 A quantum switch.

the laser pulse. Stimulated transitions are the reason why the third energy level, metastable state 1, is needed.

Each electromagnetic pulse contains many photons because transitions are probabilistic; that is, not every photon will cause a transition. When the atom is excited to state 2, it can decay to either state 1 or state 0. If it decays back to the ground state, it will be immediately re-excited by the electromagnetic pulse to state 2. In other words, while the pulse with frequency corresponding to the energy of state 2 is turned on, the only (relatively) stable state is state 1. In fact, because state 1 is metastable, the atom will remain in this state for some time after the photon source is turned off.

From a quantum computing perspective, this is not a very interesting device. Although it does have the two stable states needed for a binary logic, if we were to have many of these devices together, all we would be able to do is drive them all into one state or the other. For a computation device to be useful, it must be able to address a particular atom and change its state without affecting the other atoms. We could use different atoms each with a different excitation energy, but then we would need to have many different atoms with specific properties, and there is no guarantee that any device could be made with these atoms that might have very different chemical properties. Even if we could accomplish this feat, all we would have is a collection of one-bit processors completely uncoupled from one another.

To achieve conditional logic, that is, the ability to make the operation performed on one atom (qubit) depend upon the state of another atom (qubit), we need to understand how to bring about a coupling between any pair of atoms. This is what makes possible a conditional logic such as is needed for the controlled-NOT gate. This is shown by the energy-level "splitting" in Figure 9.2. Note how the levels of atom A change on the basis of the state of atom B (not shown). The shifts in energy level also change the energy of the photons that can effect changes in atom A.

We know that the power of quantum computers comes from their ability to manipulate superpositions and entanglements of qubits, for example, $c_0|00 \cdots 0\rangle + c_1|00 \cdots 1\rangle + \ldots$, rather than manipulating strings of classical bits. In the context of using energy levels to encode qubits, we can create and manipulate such superpositions and entanglements using pulses of laser light.

For example, there is a type of laser pulse that can change a $|0\rangle$ into a $|1\rangle$. In Chapter 1 we saw how a qubit can be pictured as an arrow contained in a sphere, with the $|0\rangle$ state represented by the

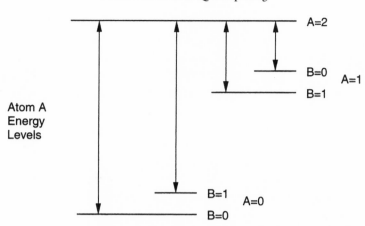

Figure 9.2   Energy level shifts can be used for conditional logic.

arrow pointing to the North pole, and the $|1\rangle$ state by the arrow pointing to the South pole. To flip a $|0\rangle$ to a $|1\rangle$, or vice versa, or in fact to change any 1-qubit state to its "polar opposite" requires a 180° change. Equivalently, measuring angles in units called radians, 180° is equivalent to π radians. So the laser pulse that flips a state to its polar opposite is called a π-pulse. But, a π-pulse is more general than a classical bit-flipper—it will flip qubits too in the sense that it will take a qubit's proportion of zeroness and oneness and reverse them. For example, an 80/20 split of $|0\rangle/|1\rangle$ would be flipped to a 20/80 split by a π-pulse. But we are not limited to these two pulse lengths. We can adjust the pulse length to be π divided by anything by changing the length of time that the laser illuminates the atom thereby creating any superposition of zeroness and oneness we desire. In particular, a $\pi/2$-pulse, which is half the length of a π-pulse, will rotate the qubit state by half of 180° (i.e., 90°). This means that if we start with a $|0\rangle$ and want to create an equally weighted superposition of $|0\rangle$ and $|1\rangle$, then we apply a $\pi/2$-pulse to the state $|0\rangle$. Thus, a $\pi/2$-pulse has the effect of converting bits into qubits.

With these tools in hand, we can now look into some particular examples of quantum gates.

Ion Traps

One of the first techniques for realizing quantum gates was devised by Juan Ignacio Cirac and Peter Zoller of the Institut für Theoretis-

Figure 9.3   Linear array of trapped ions with a laser for each ion to change its state.

che Physik at the Universität Innsbruck in Austria. This scheme uses a linear array of trapped ions as the basis for a quantum memory register as shown in Figure 9.3 (Cirac, 1995).

In the ion trap scheme the qubits are represented as the internal energy levels of ions with the ground state representing $|0\rangle$ and a metastable state representing $|1\rangle$. Thus, there is one qubit per ion (Steane, 1996). The ions are initialized for computation by means of "cooling," that is, lowering their energy to the ground state. There are many techniques for cooling ions. Some involve laser stimulation, much like the quantum computing we are trying to achieve in the first place, although using different frequencies.

Although the ions of Figure 9.3 are drawn as discrete spheres, we know from quantum mechanics that the ions are more correctly pictured as wavefunctions whose physical extent is more spread out than suggested by the sharp boundaries of a sphere. This means that the ions must be sufficiently separated so that a laser can interact, or address, each one individually. In other words, the most probable location of the adjacent ions must be at least several times the width of the laser beam used to illuminate an ion. The ions have the same electrical charge so they naturally repel one another and separate themselves in the trap. Because the ions are confined within the trap, the motion of the ions as a whole is quantized.

This quantized collective motion provides a mechanism for conditional logic. For example, when one ion absorbs a photon from a laser its outermost electron changes its quantum state and the ion itself also absorbs some momentum in the process. This momentum is then transferred to the motion of the ion, which in turn influences the collective motion of the other ions in the trap. Conditional logic is possible because now another, not necessarily adjacent, ion in the trap can be altered by absorbing this sloshing collective energy along with an appropriately designed laser pulse. The laser frequency is adjusted to match the difference between the excitation energy of the ion's electron and the collective energy frequency. This coherent motion of the ions acts as a kind of data bus that moves data back and forth along the trap. Measuring the qubits is done using a fluorescence scheme that requires an extra energy level that couples strongly to the lowest energy state. The absence or presence of the fluoresced photon tells us which state the atom was in. Under certain trap conditions, the quantum states of the ions have long coherence times so they can be used for quantum computing.

Recently, C. Monroe, D. Meekhof, B. King, W. Itano, and D. Wineland from the National Institute of Standards and Technology in Boulder, Colorado, building on the theoretical work of Cirac and Zoller, produced a working quantum logic gate that performs the controlled-NOT operation (Monroe et al, 1995). These investigators used a single beryllium ion (Be^+) in an ion trap. They were able to obtain the proper response about 90% of the time. The beryllium-based quantum gate has a decoherence time of about one millisecond.

The Boulder group was clever in encoding both of these bits within the same atom, in essence breaking the one-bit-per-atom limit. They achieved this by relying on the quantization of energy levels of the outermost electron in a beryllium ion and then putting this ion in a trap, which in turn creates an additional quantization of vibrational energy levels of the ion as a whole (rather than just that of the outermost electron of the atom). In the trap, the ion behaves as a quantum harmonic oscillator where the entire atom sloshes back and forth in the trap. The trap is made from radio frequency waves rather than from atomic barriers. The two-bits-per-atom scheme is illustrated in Figure 9.4.

The parabola represents the harmonic oscillator created by electromagnetic waves. This is the trap. The ion, shown at the center, is confined to the well because the barriers at the sides are too high for it to jump over. The height of the ion within the trap represents the

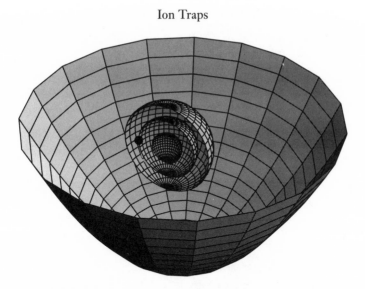

Figure 9.4 Encoding of two bits in a trapped ion.

vibrational mode of the ion. The concentric spheres represent the internal energy levels of the ion. The block dot is the energy level of the electron. By using the energy levles in the ion and the harmonic trap, we have effectively used one atom to encode two bits.

The actual details of the Boulder experiment are rather complicated, but through the judicious use of laser pulses, we can manipulate the ions and performance simple quantum logic operations. The outcomes of the operations are verified using the fluorescence technique described earlier. The absence or presence of the fluoresced photon tells us which state the atom was in.

The results of the Boulder experiment are shown in Figure 9.5. The foreground labels are the input states of the control and data qubits. The foreground towers are the four possible input states of the control and data qubits (c, d). The background towers are the results of the controlled-NOT gate operation. The black towers are the probability that the control qubit c is 1, $c = 1$. The white towers are the probability that the data qubit d is 0, $d = 0$. Note that the value of the data qubit is changed only when the control qubit is 1. If the apparatus were perfect, then all the values would be either 0 or 1. The fact that all the values are nearly 0 or nearly 1 shows that controlled-NOT can be implemented with high efficiency via qubits, even with today's technology.

One of the difficulties in scaling this particular ion trap model up to useful sizes is that each ion requires its own laser to stimulate it.

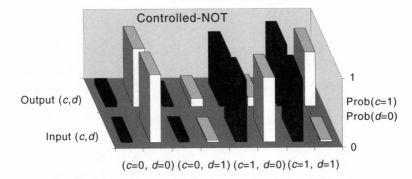

Figure 9.5 Controlled-NOT truth table results (Monroe, 1995).

Where do all these lasers come from and does the complication of so many lasers make this technology infeasible? Lasers as small as a few millionths of a meter have been around for nearly a decade (Jewell, 1991). But is even this small size too large for the millionths-of-a-meter scales typically found in quantum objects? We don't know the answers yet, but the ion trap remain one of the stronger candidate technologies. In fact, Cirac and Zoller say "We believe that the present system provides a realistic implementation of a quantum computer which can be built with present or planned technology."

In spite of these formidable obstacles, theorists continue to crank out results to tell us what is possible *in principle*. Researchers from Los Alamos National Laboratory (Hughes, 1996) have calculated that Shor's factoring algorithm could be implemented on an ion trap quantum computer (if one is ever built). So far, the largest number that could be factored would be factored using atoms of ytterbium (EE-ter-bee-um) that have a metastable state with a lifetime of 1533 *days!* With just 1926 ytterbium atoms and 30 billion laser pulses, a number with 385 bits (10^{116}, more than the number of atoms in the universe) could be factored.

"Flying Qubit"-Based Quantum Computers

Another research group at the California Institute of Technology, consisting of Q. Turchette, C. Hood, W. Lange, H. Mabuchi, and H. Kimble, has implemented a 2-qubit quantum logic gate using a very different methodology (Turchette, 1995). These investigators implement their "flying qubits" brand of quantum logic in what they refer to as a quantum-phase gate. The quantum-phase gate encodes, in the polarization states of interacting photons, information that induces a conditional dynamic in an analogous way to the models described earlier in this chapter.

In the Caltech scheme, the control and target bits of the controlled-NOT gate are implemented as two photons of different optical frequencies and polarizations passing through a low-loss optical cavity. The interaction between these photons is aided by the presence of a cesium atom drifting through the cavity, as shown in Figure 9.6.

Now the goal is to achieve a 2-qubit quantum gate that allows for the flipping of one bit conditioned on the value of another. Moreover, for this to be a truly "quantum" gate, the physical implementation must also be able to create, preserve, and manipulate superpositions of qubits and entanglements between qubits. All of these goals can be accomplished by using a cavity-based approach.

Let us imagine that the 2-qubit gate consists of a "control" bit and a "target" bit. We want to flip the target bit depending on the value of the control bit—in other words, we want a controlled-NOT.

In the cavity QED approach, the control bit is implemented as a beam of photons that either all have a "+" (left) circular polarization or all have a "−" (right) circular polarization. The intensity of the control beam is lowered so that, on average, there is only a single control photon in the cavity at any one time.

By contrast, the "target" bit is implemented as a beam of photons that all have the same *linear* polarization. Mathematically, a linearly polarized photon can be described as an equal superposition

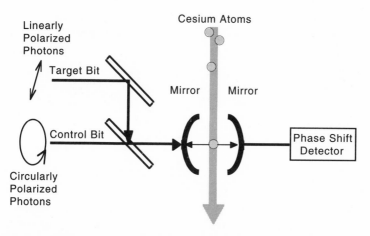

Figure 9.6 Sketch of the experimental setup for a quantum logic gate implemented using cavity QED.

of a left and a right circularly polarized state. Like the control beam, the intensity of the target beam is lowered so that there is at most one target photon in the cavity at any one time.

If the cavity were empty, the interaction between the control and target photons bouncing around inside it would be rather weak. Thus, to enhance the interaction, a cesium atom is placed in, or rather drifts through, the cavity. The spacing between the mirrors in the cavity can be tuned to resonate with a particular transition between two energy levels of the cesium atom and the target and control photons.

The Caltech researchers then measured the rotation of the "target" bit's polarization as the "control" bit's intensity was varied, when the control bit beam consisted of all + or all − photons. They observed that there is a strong correlation between the phase shift of the target photon and the intensity of the control beam only when the control photons' polarization is +. Here the linearly polarized "target" photon can be thought of as a superposition of a left and a right circularly polarized photon. When the target photon interacts with the cesium atom, the − component of its state is unchanged and the + component is phase-shifted by an amount that depends on the excitation state of the cesium atom, which, in turn, depends on whether the circular polarization of the control bit is + or −, where + states excite the atom and − states leave it in the ground state. Thus, we have a conditional transformation of the state of the target qubit depending on the control qubit.

Earlier cavity experiments by researchers L. Davidovich, A. Maali, M. Brune, J. Raimond, and S. Haroche at the Ecole Normale Superiéure, in Paris, demonstrated similar conditional phase shifts in microwave cavities (Davidovich et al., 1993). It remains to be seen whether the cavity QED techniques can be extended to complicated quantum circuits.

The ion trap and cavity QED quantum gates both assume that computation is spread over a circuit consisting of many ions all operating at high efficiency. However, in the next section we discuss a technique that packs the entire quantum computer into a single molecule.

NMR

Most of us are familiar with the fantastic advances in medical imaging. These techniques permit us to peer inside of a person's body without cutting it open or bombarding it with penetrating and destructive radiation. In many cases, patients need no longer swallow highly toxic potions to make their organs glow so that doctors can see what is happening inside their bodies. Among these new non-invasive techniques is nuclear magnetic resonance (NMR) spectroscopy, better known as magnetic resonance imaging.

Although NMR has been familiar in medical circles for over a decade, it has an even longer history in physics going back about 50 years. NMR is a highly sensitive technique for probing impurities in materials. In medical imaging, tumors can be identified because their composition is different from that of normal tissue. The spectroscopy aspect of NMR refers to the fact that the "output" of NMR is a spectrum of photons of various frequencies. By comparing a sharp-peaked spectrum against a catalog of known spectra, we can tell what is in the sample. For example, if a molecule happens to absorb a photon of a particular frequency with high probability (the resonant frequency), then we will see a dip in that frequency compared to the input photon beam. Also, if we apply certain photon pulses to an object, then there will be an emission of photons whose presence we can detect.

Earlier we spoke of using electron spins for representing qubits. Protons and neutrons also have an intrinsic spin. Starting from the simplest atom, hydrogen, and moving up the Periodic Table, a proton and typically one or more neutrons are added to each element's nucleus. As the protons and neutrons are packed in, their spins are aligned in anti-parallel fashion. If there are an even number of protons, then the contribution to nuclear spin will be zero. The same is true for the neutrons. However, if there are an odd number of protons and an even number of neutrons, then there will be a maximal nuclear spin of $1/2$, or $3/2$, or $5/2$, and so on, with larger values corresponding to larger nuclei. A particular nucleus has multiple possible spin values. For example, if the maximal nuclear spin is $5/2$, then the nucleus may also have values of $1/2$ and $3/2$. The same is true if the number of neutrons is odd and the number of protons is even. If the number of protons and the number of neutrons are both odd, the the maximal nuclear spin will be 1, or 2, or 3, depending again on the size of the nucleus. Because different isotopes of an atom have different numbers of neutrons, a particular species of atom can have different values for nuclear spin.

In order to utilize NMR in quantum computation, we first need to understand how distinct qubits can be addressed and manipulated in bulk quantity, such as in a test tube. When we say "bulk," we mean that there may be as many as on the order of 10^{23} molecules in the test tube. This is the kind of colossal redundancy that is possible with an NMR computer. As always, in order to have a useful computer, we need to address the qubits. Because all the molecules are identical, we are not actually dealing with roughly 10^{23} qubits but only with the handful of qubits for each of the 10^{23} "processors." Still, we need some method for addressing even this small number of qubits. It turns out that the energy associated with spin is inversely proportional to the mass of the particle. Thus the energy associated with nuclear spins will be roughly 1800 times smaller than the energy associated with atomic spins from electrons. This small value for the nuclear spin makes it highly sensitive—and thus ideal for our purposes. The exquisite sensitivity of each type of nucleus to its environment arises from two sources. First, the motion of the electrons in the electron cloud surrounding an atom gives rise to a magnetic field. Other than the electrons of its own atom, a nucleus is most influenced by the closest electron orbitals of neighboring atoms. The shape of these orbitals, and hence the magnetic field they generate, varies with the type of atom. These fields combine with the externally applied magnetic field of an NMR apparatus to make the field experienced by the nucleus slightly different depending on the chemical composition of the molecules in the sample. This means that the location of the resonance peaks—the anomalously large absorption peaks seen at a specific frequency—of a particular type of nucleus can shift in a way that reflects the chemical composition of the sample. This effect is known as the *chemical shift*.

In addition to chemical shifts, a nucleus can also couple with the nuclear magnetic field of neighboring nuclei, leading to an even finer structure in the NMR spectrum. The latter effect is called *spin–spin coupling*. What was initially one resonance peak can become several resonance peaks of slightly different frequencies through spin–spin coupling. Figure 9.7 illustrates the effect of spin–spin coupling on the energy levels, and hence resonance frequencies, of a 2-spin quantum memory register. The double-headed arrows illustrate the transitions allowed by quantum mechanics. These are transitions between states that differ by at most one bit—that is, one spin state. The short, single-headed arrows on the left side are the spin states of the two nuclei. In Figure 9.7 the external magnetic field is in the direction from top to bottom in the figure. An up arrow indicates that the nuclear spin is aligned opposite to the main magnetic field of the NMR device. Note that the lowest energy state occurs when the spins are aligned with the field, and the highest energy state occurs when the two spins are aligned opposite to the magnetic field.

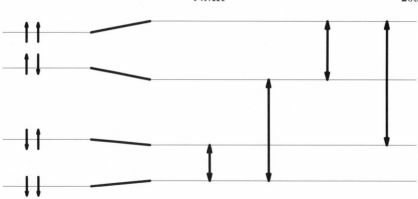

Figure 9.7 Energy-level diagram for a quantum memory register built from the spin states of two nuclei.

The left-hand side shows the energy levels that would exist if spin–spin couplings were ignored. The right-hand side shows the effect that spin–spin coupling has on the energy levels. Hence each transition can be distinguished and brought about by bathing the sample in photon pulses of appropriate frequency. Spins not sensitive to the photon frequency will not be affected. Note the similarity to the conditional logic we described earlier (see Figure 9.2).

Thus, by way of chemical shifts and spin–spin coupling, the NMR spectrum of a particular type of nucleus can look slightly different, depending on the details of its environment. In NMR spectroscopy these subtle differences in the spectra are exploited to identify the molecular structure of the sample. In NMR-based quantum computing, they enable us to use finely tuned radio frequency pulses that allow desired quantum transformations to be applied to all the nuclei at a particular location in every molecule of the sample. One of the key aspects of NMR quantum computing is that the nuclear spins *within* a molecule may be coupled but are relatively isolated from the other molecules in the sample. Thus, each molecule is essentially acting as a separate quantum memory register. We can therefore apply transformations to a qubit at a specific location in each molecule by controlling the intensity and length of radio frequency pulses. Hence, in principle, we can perform quantum computations.

In particular, we can perform the kinds of conditional logic operations on nuclear spin states that we discussed previously for atomic states (Cory, 1997). For example, to achieve the controlled-NOT operation, we need to be able to flip one bit in a pair of bits when the other bit is in the "up" spin state. We can measure which state the nuclear spin is in because the spin will precess, like a tilted spinning top, and in the process emit photons that the NMR apparatus can detect.

Figure 9.8 shows how an NMR computer can implement the controlled-NOT function using hydrogen and carbon nuclear spins (Gershefeld, 1998). There are two atoms: one carbon, C, and the other hydrogen, H. Two cases are shown. On the left side of the figure, the hydrogen is in the 1 state (the dark on the up side means 1), and on the right side of the figure, the hydrogen is in the 0 state (the dark on the down side means 0). Not shown is the external magnetic field of the NMR device, which in this case is vertically oriented. Reading the next row, an RF pulse is applied to rotate the carbon nuclear spin into a superposition of 0 and 1. While it is in this superposed state, the spin will precess around the vertical axis at a certain speed. On the next row, another RF pulse is applied, and it reverses the initial 90° rotation so that the carbon's spin is either aligned or anti-aligned with the external magnetic field. In the case on the left side, the carbon finishes in a flipped state because the hy-

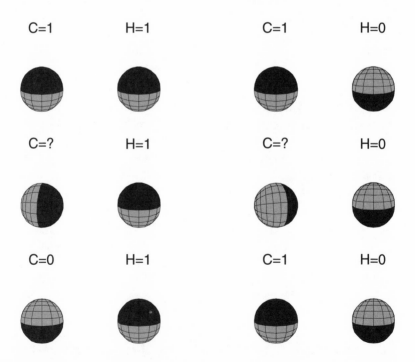

Figure 9.8 The controlled-NOT as implemented in a spin-based NMR quantum computer.

drogen was in a 1 state. In the case on the right side, the carbon finishes unchanged because the hydrogen was in a 0 state, as expected
from controlled-NOT. Why does the carbon flip in one case and not
the other? The reason is that the speed of precession depends on the
state of the hydrogen atom, which in turn determines whether the
second RF pulse has the correct resonant frequency to flip the spin.

Figure 9.9 shows a hypothetical inverted NMR spectrum of a
quantum memory register consisting of two nuclear spins before
and after the application of a sequence of radio frequency pulses
that implement the controlled-NOT operation. The top figure corresponds to the initial state, the bottom figure to the final state. The
initial NRM spectrum corresponding to the state with spin 1 is 1,
and that corresponding to the state with spin 2 is 0. The bottom figure is the NMR spectrum after the controlled-NOT operation has
been applied, corresponding to the state where spin 1 and spin 2 are
both 1. Note that the location of the peak has shifted. It is this
change in the NMR spectrum that indicates the corresponding
change in the state of the quantum memory register.

As we noted earlier, the output of the NMR quantum computer
is a spectrum of electromagnetic energy—that is, photons. The spectrum contains photons from all over the sample. We have no way of

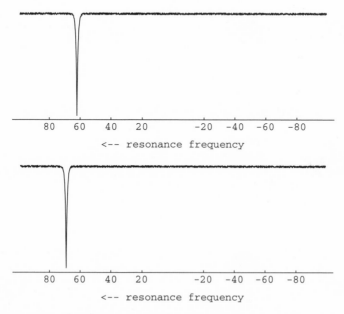

Figure 9.9 Inverted NMR spectrum of a quantum memory register consisting of
two nuclear spins.

knowing which molecules will be hit by which input photons, nor do we care. What we care about is that the spectrum is the average behavior of everything that happened in the sample material while the programming pulses were applied. It doesn't even matter if some of the molecules were corrupted by cosmic rays or contaminating chemicals, because the redundancy is so high. The output spectrum will be unambiguous (except in cases of extreme corruption of the molecules or a very bad laser), because the peaks will be so high and so narrow compared to the background that only one interpretation of the output is possible.

If it is possible to make a universal NMR-based quantum computer, it is natural to wonder how large a quantum computation can be performed in practice. Through some fairly simple calculations, we can show that the greatest possible size of the quantum memory register is roughly 73 qubits. This puts a crude limit on the size of a quantum memory register we could conceive of for a test-tube-sized quantity of sample. Using larger samples will improve this figure, of course, but the improvement grows only logarithmically with the mass of the sample. Nevertheless, this means that a quantum computer that can operate on roughly 73 qubits seems feasible, given current NMR technology. NMR quantum computation represents one extreme—ultra-high redundancy achieved by settling for fairly simple computational power.

Despite these possible restrictions on the scalability of NMR quantum computers, they do have the advantage of well-known technology behind them. This, among other considerations, led to NMR being the method of choice for the first demonstration of a quantum algorithm. This experiment implemented a simple example of Grover's quantum search algorithm (see Chapter 3). In this experiment the problem was to ask two questions concerning four numbers, a simple kind of database query. Classically, an algorithm would have to check each number, so it might have an answer to the two queries on the first, second, third, or fourth try, depending on the order in which the numbers were tried. Thus, the average number of tries before the answer is obtained is $1 + 2 + 3 + 4 = 10$ divided by 4, which is 2.25. The NMR quantum computer was able to do the search in a single try every time. For example, suppose the numbers we had to choose from were 1, 2, 3, and 4, and suppose one query was "Find an odd number" and the second query was "Find a number greater than 2." The only number of the four that satisfies those two queries is the number 3. The NMR quantum computer can effectively ask the queries of all four numbers at the same time by virtue of using superposition. Thus, it takes only one try to find the answer.

The quantum database query experiment was done by Isaac Chuang of IBM, Neil Gershenfeld of M.I.T., and Mark Kubinec of the University of California at Berkeley (Chuang, 1998). The team used the nuclear spin of a hydrogen nucleus (a proton) and a carbon nucleus inside of a chloroform molecule, $CHCl_3$. The experimenters operated their quantum computer at room temperature and used a sample of only half a milliliter (about one-eighth of a teaspoon), but that nonetheless contained about 10^{20} chloroform atoms—a hundred million trillion quantum computers! The steps involved in the algorithm follow the same "prepare, evolve, measure" sequence that we have been discussing all along.

Specifically, the first step was to prepare the spins in each chloroform molecule into a superposition. This was done by placing the sample inside the powerful magnetic field produced by NMR coils. The coils had a strength that was more than 200,000 times stronger than the magnetic field at Earth's surface and was homogeneous to one part in a billion—in other words, don't try this at home. The hydrogen atoms and the carbon atoms (the carbon-13 isotope in this case) have different masses, which means that they respond to different ratio frequencies. Thus, either atom can be "addressed" by the use of a particular resonant frequency. Thus, the *prepare* step involves an application of two RF pulses of different frequencies and of a certain duration to put the hydrogen and carbon atoms of each chloroform molecule into a superposition. The hydrogen atom was used as the "input/output" and the carbon as the "working" qubit, where the answer is eventually found. Next, an RF pulse of a particular frequency rotates the nuclear spin to a direction perpendicular to the main magnetic field of the NMR coils. In this intermediate position between being aligned and being anti-aligned, the spins are effectively in a superposition.

The next step in the process is the evolution of the algorithm. Once again, the computer is driven by the application of RF pulses. Now what happens is that these pulses induce a spin evolution to occur in the nuclear spins of the hydrogen and carbon atoms. As always with quantum computing, we must be concerned with the coherence time of the operations. In this experiment the longest computation took just 35 thousandths of a second, which was much less than the coherence time of a few tenths of second. At this point, because the input qubit is in a superposition of all possible inputs, we have essentially also put the second qubit into a superposition as well, thus realizing quantum parallelism.

Finally, we must "read" the answer from our computation by a measurement. First we must apply more RF pulses to bring the spins back to their orientation to the main magnetic field. The actual answer to the problem is in the first qubit, the hydrogen atom. The state of the hydrogen atoms in the chloroform sample is measured with a wire coil surrounding the apparatus. A voltage is induced in the coil by the precessing of the spins of the hydrogen nucleus (a single proton) in the main magnetic field. The voltage spectrum is converted into a frequency spectrum. Whether the peak in the spectrum is up or down determines whether the atom was in a 0 or a 1 state. Without going into the details, let us simply say that the experimenters seemed to confirm that their apparatus did indeed perform the Grover algorithm, and the results were as predicted by the theory, within the uncertainties of the experimental setup.

This group did not stop there, however, and a second quantum computing program was attempted. In fact, two independent groups attempted to implement this algorithm. One group comprised Isaac Chuang again, Lieven Vandersypen, Xinlan Zhou, and Debbie Leung from Stanford, and Seth Lloyd from M.I.T. (Chuang et al., 1998). The other group consisted of J. Jones and M. Mosca from Oxford; the Oxford group used cytosine, $C_4H_5N_3O$, rather than chloroform for its quantum computer (Jones and Mosca, 1998). These groups seemed to demonstrate not only a true quantum algorithm but quantum parallelism as well. The particular algorithm they attempted to implement was the Deutsch–Jozsa algorithm mentioned in Chapter 2. Recall that in this algorithm we are trying to determine whether a mathematical function that has a 1-bit output is a constant or has an equal number of zero and one outputs based on all its possible inputs.

The simplest form of the Deutsch–Jozsa algorithm involves two qubits. For the purposes of this book, we will focus on what the IBM–Stanford–M.I.T. team implemented. Classically, this algorithm requires two checks of the function, whereas the quantum case requires only one.

In this experiment the coherence times to perform the various operations were about 20 seconds for forming the superposition and about a third of a second for the actual algorithm—both far longer than the 7 thousandths of a second required to perform the computation. This might seem like an eternity simply to manipulate two bits, but in fact a classical computer could not perform the Deutsch–Jozsa algorithm with the same economy of steps. Using the same techniques as before, the experimenters believed they had confirmed that their apparatus did indeed perform the Deutsch–Jozsa algorithm using quantum parallelism.

At this state of quantum computing research, we just want to demonstrate that the theoretical principles are realizable; the issue of optimizing performance comes later. The demonstration of actual quantum algorithms on specialized quantum computers is certainly encouraging, but as we pointed out earlier, some serious issues arise in the scaling up of NMR.

The attentive reader will have noticed the carefully worded phrase "seemed to" in the previous descriptions of implementations of quantum parallelism. The reason for this is that in the months following the initial NMR experiments, several theorists questioned whether the NMR experiments were capable of producing truly entangled states (Braunstein et al., 1998). At the time of writing this book, the consensus is that NMR techniques with only a few qubits are incapable of producing true entanglement between the qubits. This means that NMR quantum computing might be more limited than had hitherto been believed and might only be applicable to quantum algorithms that do not reply upon entanglement. However, NMR techniques might still be used to investigate certain quantum physical effects.

The Kane Mutiny

So far our discussion of potential embodiments of quantum computers has relied upon novel, if not downright exotic, physical phenomena. We should ask ourselves: Is it possible to bootstrap off the enormous investment in silicon-based computer technology to make quantum computers? If so, it might offer the fastest possible route to practical quantum computation. Indeed, it seems that it may be possible, as we now describe.

The design for a silicon-based quantum computer was devised by Australian physicist Bruce Kane (Kane, 1998). Kane's scheme utilizes so-called *hyperfine* interactions, which couple nuclear spins to electron spins. The qubits are stored in nuclear spins but are made to interact indirectly via a coupling mediated by electron spins. When we have a way of making the spin state of one nucleus influence that of another nucleus, we have the possibility of performing the conditional quantum logic—and hence general quantum computation.

The architecture for Kane's silicon-based quantum computer is shown in Figure 9.10. It consists of a layer of silicon mounted on a substrate (which is not shown explicitly). Within the silicon, placed at regular intervals, are phosphorus atoms that each sit beneath a metal strip, called an "A-gate." Phosphorus is the element that comes after silicon in the periodic table. The most common isotope of phosphorus is called ^{31}P. This means that there are 15 protons and 16 neutrons in each nucleus of a ^{31}P atom. The fifteenth, or "last," proton is the one whose spin is used as a qubit.

When placed in a low temperature environment—0.1 Kelvin— the outermost electron (e^-) is only loosely bound to its host phosphorus atom. In Figure 9.10 this is shown by the extended gray cloud, which represents the probability density of where the electron may be found. In this weakly bound state the electron's spin can influence the state of the phosphorus nuclear spin.

"A-gates" are metal strips bonded to a barrier separating the gates from the silicon. By applying a voltage to an A-gate we can distort the probability distribution of the electron position, as shown by the rightmost electron in the top of the figure. In particular, this distortion allows us to control the strength of the hyperfine

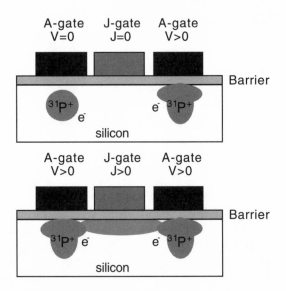

Figure 9.10 Proposed silicon-based array architecture for a quantum computer.

interactions between the electron and nucleus. This in turn alters the resonant frequency of the nuclear spins in the phosphorus nuclei. Hence, by applying a tiny voltage to an A-gate we can tune any phosphorus nucleus to react to a slightly different frequency of applied oscillating magnetic field. This allows us to operate selectively on specific qubits in the array of phosphorus atoms.

"J-gates" have the effect of turning on or off the electron-mediated coupling between adjacent nuclear spins. Together, the A- and J-gates allow us to prepare, evolve, and measure the nuclear spins. These types of gates also permit us to address each of these spins individually by making use of the coupling of each spin with its neighbor's spin [for an analogous effect see (Lloyd, 1993)].

Thus, the "working" spins of the Kane design are the nuclear spins, while the electron spins act as coupling agents between the nuclear spins. The electrons affect the nuclear spins when they are "close" to one another in a probabilistic sense. This closeness is controlled by the J-gate which, when turned on, acts as a "probability bridge" between donor electrons from neighboring phosphorus atoms.

The electron spins also act as coupling agents for the measurement process. In other words, although the nuclear spins store the actual qubit values, we obtain the answer by coupling, that is, effectively transferring, the nuclear spins to the electron spins. This in turn affects the electron's probability distribution, which can be examined with charge-sensitive measurements among the A-gates. If a different voltage is applied to the A-gates and the electrons are in a low-energy state, then there will be a small current due to the motion of one of the electrons as it moves between phosphorous ion donors that can be measured. More specifically, the electron spins are usually aligned with the strong external magnetic field. However, when the J-gate voltage is changed appropriately, then the more likely configuration of the electrons is to be anti-aligned and the aligned state becomes only metastable. But, whether the anti-alignment occurs depends upon the direction of the ^{31}P spin that is most strongly coupled to its electron. It then becomes possible for one of the phosphorus ions to bind both electrons, but only if the electron spins are anti-aligned; otherwise, there is no electron migration. The hopping of an electron from one ^{31}P to the other creates a current that can be measured on the A-gates. In summary, it is the various states of the nuclear and electron spins and the strength of the gates that determines whether an electron will hop or not and this is what we ultimately measure as part of the quantum computation.

As with any candidate for a quantum computer we need to ask about the decoherence time. Fortunately, the decoherence time for the electron motion is long—thousands of seconds. The nuclear spin decoheres even more slowly—roughly 100 billion *years*.

While Kane's silicon-based quantum computer relies on materials similar to today's machines, there are significant fabrication challenges. His design envisions devices more than a hundred times smaller than today's mass-produced chips. If extrapolations to chip sizes hold, then it will be years before such sizes are commonplace. Still, that puts it right on the path toward the 2020 limit.

Summary

Table 9.1 summarizes the principles on which the four embodiments of quantum computers discussed in this chapter, that is, ion traps, cavity QED, NMR, and silicon, are imagined to work. To be a plausible architecture for a quantum computer, any candidate embodiment must be able to demonstrate certain core capabilities. These include the ability to make a set of qubits, to initialize those qubits in some desired starting state, to perform any 1-qubit gate operation on each qubit, to perform a controlled-NOT operation between any pair of qubits, to readout the bit value of any qubit, and to keep the whole system coherent long enough for a desired computation to be completed. Table 9.1 summarizes how each of these goals is achieved in the context of the various embodiments considered in this chapter.

Whichever scheme turns out to be the most feasible, the advances made along the way will no doubt have useful spin-offs—in basic physics research at the very least. For example, being able to create and manipulate just a handful of qubits allows us to run some interesting tests on certain predictions of quantum theory. Thus, ironically, quantum computing may eventually become a tool of experimental physics. As we have seen, only a small number of qubits are needed to simulate quantum systems much more efficiently than can be done classically. Looking ahead, at around the 10-qubit level, a quantum computer will be able to implement certain coding schemes for fault-tolerant computing. At around the 100-qubit level a quantum computer could be used as a repeater in a noisy quantum cryptographic channel. And at the level of a few thousand qubits, a quantum computer can factor large integers.

Table 9.1
Requirements for a quantum computer architecture and their embodiments.

Achieved	Ion Trap	Cavity QED	NMR	Silicon
Qubit	Energy levels in an ion	Polarization states of a photon	Spin states of a nucleus	Spin states of ^{31}P nucleus
Preparation	Ion cooling	Preparing linearly and circularly polarized photons	Setting the average state of spins in the sample	Applying voltages to individual ions
Evolution	Applying laser pulses of specific frequencies	Photon–photon interaction	Applying radio frequency pulses of specific frequencies	Applying voltages to voltage-controlled gates
Conditional logic	Coupled collective vibrations of trapped ions	State of cesium ion and photon polarization	Nuclear spin–spin interactions in same molecule	Coupling between atomic and nuclear spins
Readout	Resonant absorption and fluorescence	Phase shift detector	Absorption spectrum	Voltage measurements between electrodes
Coherence	Well-isolated	Well-isolated	Using pseudo-pure states in a decoherent environment	Well-isolated

Thus, experimental quantum computing holds much promise long before the goal of a general-purpose quantum computer has been reached.

Ten

⤐◈⤏

It Is Now Safe to Turn Off Your Quantum Computer

The most beautiful thing we can experience is the mysterious.
—Albert Einstein

By now you will realize that many of the assumptions we make about classical computers cease to be correct at the quantum scale. We have seen, for example, that a quantum bit is not necessarily a 0 or a 1, but can be a superposition of both 0 and 1 simultaneously. We have seen that a quantum bit does not necessarily have a definite bit-value until the moment after it has been read. We have seen that reading one qubit can have an effect on the value of another, unread qubit, if the two qubits are initially entangled with each other.

We thought we'd finish the book with perhaps the most unusual property of quantum computing so far discovered. This is the realization that a quantum computer can answer certain questions by virtue of the fact that it has the *potential* to answer such questions. It is not always necessary to actually run the computer to obtain the answer. The mere possibility that the quantum computer *would* give the correct answer, *if* it were run, is sufficient to deduce the answer!

Such a claim might sound crazy. How can an event that does not happen cause a subsequent event? In the theory of logic, however, such events are called "counterfactuals" because they run counter to the facts of the situation. In the case of quantum counterfactuals we have a situation where some sort of measurement *could* have been made but was *not*, yet we gain some information nonetheless. In the words of Roger Penrose *"What is particularly curious about quantum theory is that there can be actual physical effects arising from what philosophers refer to as counterfactuals—that is, things that might have happened, although they did not in fact happen"* (Penrose, 1994). In other words, whereas classical counterfactuals do not have tangible physical consequences, quantum counterfactuals do. The mere possibility that some quantum event might have happened changes the probabilities of obtaining various experimental outcomes.

One context in which quantum counterfactuals arise is in the area of "interaction-free" measurement. An interaction-free measurement is a measurement in which no energy is exchanged between the probe and the object, but the presence of the object is detected anyway. The fact that no energy is exchanged does not preclude the possibility of other quantum numbers being exchanged, so the term "interaction-free" is a slight misnomer. For interaction-free measurement to work, however, it is essential that the object has the potential to absorb energy from the probe, but in fact does not do so.

The idea behind interaction-free measurement goes back to 1960 when Renninger described the notion of a "negative-result measurement" which characterized a particular type of nonobservance as a kind of nondisturbing quantum measurement (Kwiat et al, 1995). Later, in 1981, Dicke described a system in which an atom is deduced to not be located in a photon beam based on the absence of scattered photons from the beam (Dicke, 1981). More recently, Avshalom Elitzur and Lev Vaidman of Tel-Aviv University proposed a more cunning means of interaction-free measurement based on properties of the photons passing through quantum interferometers (Elitzur and Vaidman, 1993).

Quantum Interferometry: It's All Done with Mirrors!

An interferometer is an arrangement of optical elements such as beam splitters, mirrors, phase-shifters and photon detectors. When a photon interacts with one of these optical elements, its quantum state is transformed in some way. For example, if a single photon is fed into one port of a lossless symmetric beam splitter, the output quan-

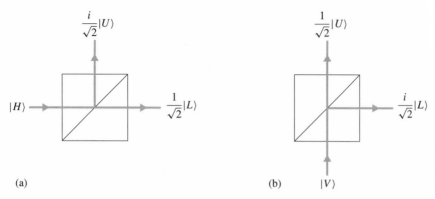

Figure 10.1 Action of a lossless symmetric beam splitter on an incident photon when the photon enters (a) the horizontal port or (b) the vertical port of the beam splitter.

tum state of that photon becomes an equally weighted superposition of a reflected photon and a transmitted photon. In Figure 10.1(a) we see an incident photon enter a beam splitter via the horizontal port and in Figure 10.1(b) via the vertical port. Either way, the output photon is left in a superposition of a transmitted and reflected component with a relative phase shift of 90° between them. The factor of $i = \sqrt{-1}$ arises because a phase shift of $\theta = 90°$ modifies the amplitude by the factor $e^{i\theta} = \cos\theta + i\sin\theta$, which equals i when $\theta = 90°$.

A particular example on an interferometer is the Mach–Zehnder interferometer shown in Figure 10.2. This interferometer can be configured so that any photon entering the horizontal input port of the lower left beam splitter is *guaranteed* to exit through the horizontal output port of the upper right beam splitter.

A single incoming photon in state $|H\rangle$ enters the interferometer via the *Horizontal* input port of the beam splitter, BS1, in the lower left corner of Figure 10.2. When the photon encounters the beam splitter there is a 50% chance that the photon will pass straight through and proceed along the lower path, L, and a 50% chance it will be reflected (acquiring a 90° phase shift in the process) and proceed along the upper path, U. Thus, the beam splitter, BS1, has the effect of creating a superposition of a photon travelling along the upper path and the lower path simultaneously. As the probabilities of transmission and reflection are equal, each component of the photon is weighted by a factor of $1/\sqrt{2}$. The factor of i indicates that there is a 90 degree relative phase between the components of the photon in the upper and lower paths.

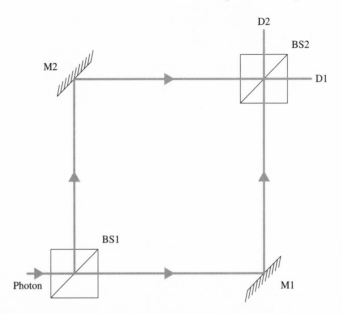

Figure 10.2 A Mach–Zehnder interferometer set up so that a single input photon is guaranteed to exit via the horizontal output port of the upper right beam splitter and be registered in detector D1.

$$|H\rangle \xrightarrow{\text{BS1}} \frac{1}{\sqrt{2}}|L\rangle + \frac{i}{\sqrt{2}}|U\rangle$$

Regardless of which path taken, the photon is reflected off a mirror (M1 or M2). Each mirror gives an additional phase shift of 180° to the photon, thereby changing its quantum state as follows:

$$\frac{1}{\sqrt{2}}|L\rangle + \frac{i}{\sqrt{2}}|U\rangle \xrightarrow{\text{M1/M2}} -\frac{1}{\sqrt{2}}|L\rangle - \frac{i}{\sqrt{2}}|U\rangle$$

The upper and lower paths are then recombined at beam splitter BS2. Quantum-mechanical interference between the components of the photon in the upper and lower paths will occur whenever we cannot tell which path the photon actually took. As there is no way to extract this information from the interferometer shown in Figure 10.2, the photon will self-interfere at BS2. If the lengths of the upper and lower paths are adjusted so that they are exactly equal, this self-interference will result in 100% certainty that the photon will be registered in detector D1 and *never* in detector D2.

$$-\frac{1}{\sqrt{2}}|L\rangle - \frac{i}{\sqrt{2}}|U\rangle \xrightarrow{\text{BS2}} -\frac{1}{\sqrt{2}}\underbrace{\left(\frac{i|H\rangle + |V\rangle}{\sqrt{2}}\right)}_{\text{from lower path}} - \frac{i}{\sqrt{2}}\underbrace{\left(\frac{|H\rangle + i|V\rangle}{\sqrt{2}}\right)}_{\text{from upper path}}$$

$$= -i|H\rangle$$

In other words, the contributions of the photon in the upper and lower paths constructively interfere in the horizontal output port of BS2 and destructively interfere in the vertical output port of BS2. Provided the lengths of the two arms of the interferometer are exactly equal, and provided the beam splitters and mirrors act perfectly, an input photon can *only* emerge from the horizontal port of the final beam splitter and *never* from its vertical port.

Quantum Bomb-Testing

Now, you might think that such an interferometer is a rather boring device. All we seem to have done is put a photon in one end and watch it come out the other. It turns out, however, that Mach–Zehnder interferometers are remarkably versatile. For example, if we place one in a rotating reference frame, the rotation induces an additional phase difference between the photons in the two arms of the interferometer, which in turn creates a new pattern of correlations between the firing of detectors D1 and D2 that allows us to infer the rotation rate. In other words, we can make a gyroscope that has no moving parts! Likewise, if instead of an optical interferometer we use a vertically aligned matter-wave interferometer that uses atoms instead of photons, the difference in gravitational fields in the two arms of the device gives rise to a phase shift and hence pattern of detection events between the two output detectors from which we can infer the gravitational gradient. Such a device is a quantum gravity gradiometer. In both cases the sensitivities of these instruments can be made thousands to millions of times better than anything that is possible classically. So as you can see, with a few minor tweaks we can mold a Mach–Zehnder interferometer into a highly versatile scientific instrument.

One of the most surprising applications for a Mach–Zehnder interferometer is as an interaction-free measurement device. Such an application was given a dramatic spin by Elitzur and Vaidman when they imagined using it for detecting an ultrasensitive bomb. The scenario goes like this: Imagine a Mach–Zehnder interferometer set up as shown in Figure 10.2. In such a circumstance a photon entering the device will certainly exit at the horizontal port of the output beam splitter. But now imagine what happens if someone places a bomb in the lower arm of the interferometer. We imagine that the bomb is a perfect absorber so that if just a single photon falls on it, the bomb will explode. The question is, Is it possible to detect the presence of the bomb without triggering it?

On the face of it, this seems like an impossible task. Common sense dictates that to see something you have to illuminate it and then detect the reflected or scattered light. The very least amount of light you could use would be one photon's worth. But in the strange world of counterfactual quantum mechanics there are ways to infer the presence of an object without actually seeing it.

Consider what happens to a single photon passing through a Mach–Zehnder interferometer that does indeed contain an ultrasensitive bomb in one arm of the interferometer as shown in Figure 10.3.

When the bomb is blocking one arm of the interferometer, with probability 1/2 the photon will be transmitted at the first beam split-

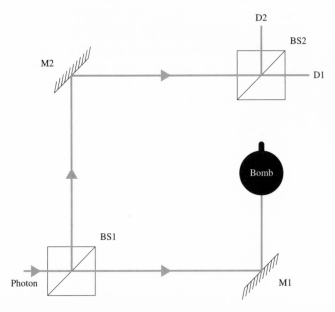

Figure 10.3 A Mach–Zehnder interferometer set up so that a single input photon can either strike the bomb, causing it to explode, or avoid the bomb altogether and emerge from either the horizontal or vertical ports of the second beam splitter. If the photon emerges from the vertical port and is detected by detector D2, this proves the bomb is present even though the photon did not touch it.

ter, travel along the lower arm (containing the bomb), and therefore cause the bomb to explode. However, also with probability 1/2, the photon will be reflected at the first beam splitter, travel along the upper arm and be registered in either detector D1 or D2 (with equal probability). Now if the photon is registered by D1, we cannot infer whether or not the bomb is present. This is because if the bomb were absent the photon would certainly be detected by D1, but also if the bomb is present the photon has a probability of 1/4 of being detected by D1, too. However, if the photon is detected by D2 (an event that occurs with probability 1/4 overall) then we can conclude that the bomb must be present even though no photon touched it! For the photon to be detected by D2, that photon must have traveled along the upper arm of the interferometer and hence could not have touched the bomb. The probabilities for the various possible outcomes are shown in Table 10.1.

Instead, of placing the bomb *in* the path of the photon, it is actually sufficient to couple the bomb to a sensor that measures whether or not the photon travels along a particular path. The sensor performs what is called a "quantum nondemolition" (QND) measurement on the photon, meaning that the sensor can register the passage of the photon without disturbing the observable property of the photon that we care to monitor. This allows the photon to proceed to the beam splitter BS2 rather than being absorbed by the bomb.

Thus, if an ultrasensitive bomb lies in one arm of the interferometer, we have a probability of 1/4 of detecting it without causing it to explode. This also means that, if a bomb is present we have a probability of 1/2 of causing the bomb to explode. This is an undesirable outcome, so a natural question is whether we can boost the probability of detecting the bomb without exploding it?

Table 10.1

Probabilities of various events in an interaction-free measurement.

Probability	Detection Scheme	Was Bomb Touched?	Is Bomb Present?
1/2	Photon is absorbed by the bomb and the bomb explodes.	Yes!	Yes!
1/4	Photon registered in detector D1.	No!	Inconclusive
1/4	Photon registered in detector D2.	No!	Yes!

Indeed we can. In fact, it is possible to boost the probability of making an interaction-free measurement arbitrarily close to 1 (Figure 10.4). One way to do this is to combine the Elitzur–Vaidman scheme with the quantum Zeno effect. The quantum Zeno effect is the ability to freeze a quantum system in a certain state by subjecting the system to a rapid succession of identical measurements. Each measurement has the effect of projecting the system back into its initial state, thereby preventing it from evolving into a new state. For example, suppose a horizontally polarized photon is made to pass through a sequence of N polarization rotators. Each rotator tends to tilt the polarization of the photon through say $(90/N)°$. Thus, after passing through all N rotators, the polarization of the photon will be tipped a full 90° from horizontal to vertical.

To use the quantum Zeno effect to boost the probability of success of interaction-free measurement we use a probe photon pre-

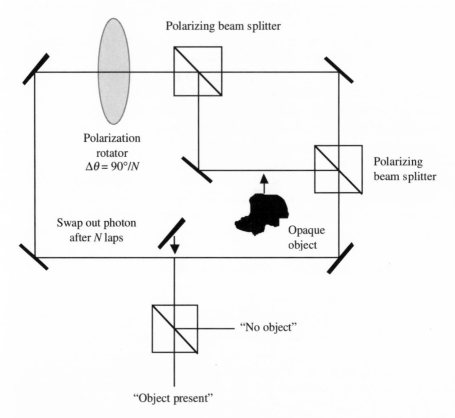

Figure 10.4 Scheme for boosting probability of making an interaction-free measurement. (Adapted from Kwiat et al., 1999.)

pared initially in a horizontal polarization state. In the absence of an object (or bomb) the photon is made to circulate N times around an interferometer. On each pass the polarization state of the photon is made to rotate by $(90/N)°$. After N laps of the interferometer the photon is swapped out of the loop and its polarization state is analyzed. If an object is present in one arm of the inner interferometer, the rotation of the polarization of the photon is inhibited. When the photon is swapped out of the loop and examined, if it still has its original (horizontal) polarization, this means that an object must have been present. If the polarization is vertical, no object was present. The efficiency of the interaction-free measurement scheme can be made arbitrarily close to one by selecting a sufficiently large value for N (Kwiat et al., 1999).

Counterfactual Computing: Computing Without Computing

The trick underpinning interaction-free measurement can be exploited in the realm of quantum computation too. Suppose we build a quantum computer capable of answering a certain decision problem, such as deciding whether a given number is prime (i.e., divisible only by itself and unity). We further imagine that the computer would operate flawlessly if it were run. Thus, if given the number 17 as input, our computer would give the correct decision, namely, that 17 is prime. We can record this decision in a memory register $|r\rangle$ that is set equal to $|r = 1\rangle$ if the number is indeed prime and left in its initial state of $|r = 0\rangle$ otherwise.

We make the switch of the quantum computer sensitive to light so that, if even a single photon touches the computer, it performs its computation; otherwise, it remains in its initial state and does not perform any computation at all. The miracle of quantum computing is that the mere fact that the computer *would* give the correct answer *if* it were run, is enough for us to infer the answer even though the computer is, in fact, *not run*. This remarkable re-interpretation of the Elitzur–Vaidman bomb was first proposed by physicist Richard Jozsa of the University of Plymouth (Jozsa, 1998).

Jozsa's thought experiment envisions using a Mach–Zehnder interferometer that contains the quantum computer in one arm of the interferometer. The "on/off" switch of the computer is the presence or absence of a photon in the arm containing the quantum computer (see Figure 10.5).

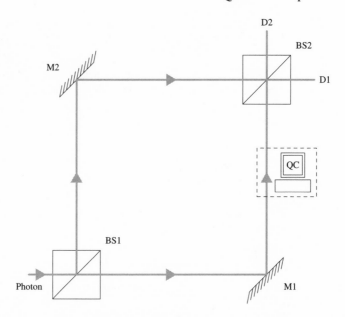

Figure 10.5 A Mach–Zehnder interferometer used to illustrate counterfactual quantum computing.

If the photon passes by the quantum computer, the QND sensor is triggered and the computer performs its computation. This entails running the primality test algorithm on some prearranged number, recording the decision in an answer register, and then resetting the quantum computer to its initial state. Otherwise, if the photon travels along the arm that does not contain the computer, then no computation is performed and the quantum computer remains in its initial state.

To trace through the steps in counterfactual quantum computing works, we will describe the quantum state of the on/off switch, the computer, and the memory register containing the result as $|switch\rangle|computer\rangle|register\rangle$.

We begin by placing the switch in a superposition of on and off simultaneously. To do this we require a Walsh–Hadamard gate, as we saw in Chapter 2. To make such a gate out of optical elements, we need to supplement our lossless symmetric beam splitter with two $-\pi/2$ phase shifters as shown in Figure 10.6 (Cerf, Adami, Kwiat, 1997).

The Walsh–Hadamard gate, performs the following transformations:

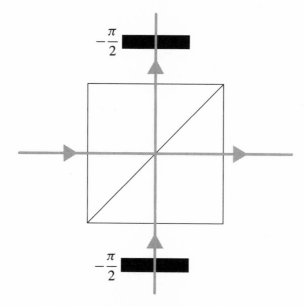

Figure 10.6 Lossless, symmetric beam splitter retrofitted with two phase shifters converts the beam splitter into a Walsh–Hadamard gate.

$$|off\rangle \rightarrow \frac{1}{\sqrt{2}}(|off\rangle + |on\rangle)$$

$$|on\rangle \rightarrow \frac{1}{\sqrt{2}}(|off\rangle - |on\rangle)$$

Mathematically, this is equivalent to the operator $\hat{W} = \left(\begin{smallmatrix} 1/\sqrt{2} & 1/\sqrt{2} \\ 1/\sqrt{2} & -1/\sqrt{2} \end{smallmatrix}\right)$.

Initially, let's suppose that a single photon, representing a switch in the $|off\rangle$ state, is fed into the horizontal port of the first beam splitter. At this same moment, the quantum computer is in some initial starting state, $|comp\rangle$, and the answer register has been set to contain $|r = 0\rangle$. After passing through the first retrofitted beam splitter, we therefore create the state:

$$\left(\frac{|off\rangle + |on\rangle}{\sqrt{2}}\right)|comp\rangle|0\rangle$$

That is, the switch is in a superposition of being $|off\rangle$ and $|on\rangle$ simultaneously, the computer is in some initial state, and the answer register contains the value 0.

Next, we allow the computer enough time to complete its computation, *if* it were to run. It is important, after the computation has been completed, that the computer be returned to the state, $|comp\rangle$, in which it started out. As a quantum computation is reversible, this is always possible. Hence, after waiting the required time, the state of the switch, the computer, and the answer register become:

$$\frac{1}{\sqrt{2}}\big(|off\rangle|comp\rangle|0\rangle + |on\rangle|comp\rangle|r\rangle\big)$$

Notice that there is now entanglement between which path the photon took and the contents of the answer register.

Next we apply a final Walsh–Hadamard gate (implemented in the upper right retrofitted beam splitter) to the switch. This yields the state:

$$\frac{1}{\sqrt{2}}\left(\frac{|off\rangle + |on\rangle}{\sqrt{2}}|comp\rangle|0\rangle + \frac{|off\rangle - |on\rangle}{\sqrt{2}}|comp\rangle|r\rangle\right)$$

$$= \frac{1}{\sqrt{2}}\left(|off\rangle\frac{|0\rangle + |r\rangle}{\sqrt{2}} + |on\rangle\frac{|0\rangle - |r\rangle}{\sqrt{2}}\right)|comp\rangle$$

In this equation $r = 0$ or 1 according to whether the answer to the decision problem is "no" (the number is not prime) or "yes" (the number is prime). But until the computer is run, this answer is unknown.

Consider the possible pairs of outcomes obtained by measuring which port the photon exits the interferometer together with the corresponding contents of the answer register. The probabilities of the various possible pairs of measurement outcomes are shown in Table 10.2. The key variables are the state of the switch ($|on\rangle$ or $|off\rangle$) and the value contained in the answer register ($|r = 0\rangle$ or $|r = 1\rangle$).

The outputs we obtain will depend upon both whether the number being tested, N, is or is not prime and the path taken by photon through the device. Table 10.2 shows the probabilities of all possible pairs of observed outcomes in counterfactual computing together with their associated interpretations. There are two cases to consider: N is *not* prime and N *is* prime. Let us begin by considering what happens if N is not a prime number.

The answer register originally contains the value $|r = 0\rangle$. If N is not prime, then r will remain zero regardless of whether or not the

Table 10.2
Table of counterfactual primality testing.

Case	Probability	Switch	Register	Computer Ran in Our Universe?	Conclusion
N is not prime	1	$|off\rangle$	$|r = 0\rangle$	Unknowable	Inconclusive
	1/4	$|off\rangle$	$|r = 0\rangle$	No	Inconclusive
N is prime	1/4	$|off\rangle$	$|r = 1\rangle$	Yes[a]	N is prime
	1/4	$|on\rangle$	$|r = 1\rangle$	Yes[b]	N is prime
	1/4	$|on\rangle$	$|r = 0\rangle$	No[c]	N is prime

[a] Because r has changed to $r = 1$.
[b] Because r has changed to $r = 1$.
[c] Otherwise r would have been set equal to $r = 1$.

computer is run. Thus, if the number is not prime, there will be no way of telling which path the photon took through the device, the photon's self-interference will be preserved, and so the photon will always be found to exit via the horizontal port of the upper right beam splitter, BS2, 100% of the time. Thus, when the number being tested is not prime we will always find the switch to be $|off\rangle$, and always find the register to be $|r = 0\rangle$. Unfortunately this state of affairs also occurs 25% of the time when the number *is* prime but the computer is not run (see below). So observing the $|off\rangle|r = 0\rangle$ output does not prove conclusively that the number being tested is not prime.

Next consider what the outputs can be when the number being tested is prime. As before, the answer register starts out in the state $|r = 0\rangle$. But this time if the computer is run the answer register will be changed to $|r = 1\rangle$ but if the computer is not run the answer register will remain in the state $|r = 0\rangle$.

According to the equation for the final state of the device, i.e.,

$$\frac{1}{\sqrt{2}}\left(|off\rangle \frac{|0\rangle + |r\rangle}{\sqrt{2}} + |on\rangle \frac{|0\rangle - |r\rangle}{\sqrt{2}}\right)|comp\rangle$$

if N is prime and the computer runs, there is an equal chance of the switch being found to be $|on\rangle$ or $|off\rangle$. But if N is prime and the computer does not run, the value of r will remain at $|r = 0\rangle$, and so the switch will only ever be found to be $|off\rangle$. Hence, there are four pairs of values for the switch and the register in the case when N is prime: $|off\rangle|r = 0\rangle$, $|off\rangle|r = 1\rangle$, $|on\rangle|r = 0\rangle$, and $|on\rangle|r = 1\rangle$, each of which occurs with probability equal to 1/4 (i.e., 25% of the time). It is worth considering how to interpret each of these possible pairs of outputs.

To obtain the result $|off\rangle|r = 0\rangle$ (with probability 1/4) the computer must never have run, for if it did the answer register would certainly contain the result $|r = 1\rangle$. Remember, we are considering the case in which N is prime and we assume our computer always works perfectly. Unfortunately, even though N is prime, the result $|off\rangle|r = 0\rangle$ is identical to the result we obtain 100% of the time when N is not prime. Hence, the result is inconclusive and we can infer nothing about the primality of N.

The second pair of outputs that we might measure is $|off\rangle|r = 1\rangle$. This occurs with probability 1/4 and arises when the computer did run and the number is prime. Some people think it odd to find the switch to be $|off\rangle$ and yet $|r = 1\rangle$, but really there is no mystery here. The photon simply exited BS1 via the lower path and switched the computer on; the computer performed the primality test and recorded the result $|r = 1\rangle$ in the answer register. This record of $|r = 1\rangle$ allows us to infer which path the photon took through the interferometer. Once we have "which path" information the interference at the output beam splitter BS2 is destroyed and so the photon is then equally likely to exit either the horizontal or vertical port. When the photon happens to exit via the horizontal port it is labeled as an "$|off\rangle$" value for the switch. But really "on" and "off" only have the intuitive meaning inside the interferometer.

The third possibility, $|on\rangle|r = 1\rangle$ is akin to the "classical" result. The computer is turned on, does the computation and changes the register to signal its result. As in the case of the $|off\rangle|r = 1\rangle$ result the "which path" information destroys the interference at BS2 and permits the photon to exit via either output port. In the case of the $|on\rangle|r = 1\rangle$ result the photon happened to exit via the vertical output port of BS2.

The final possibility, that of observing $|on\rangle|r = 0\rangle$, is rather peculiar, however. In this case the photon is seen to emerge from the vertical output port of BS2 and yet the value of the answer register is still the same as its initial value, namely, $|r = 0\rangle$. Recall that we know that if the computer is run on a number that is prime it will surely decide correctly that N is prime and will set $|r = 1\rangle$. So to find $|r = 0\rangle$ and yet (by knowledge of the case we are considering) know that N *is* prime, we must infer that the computer did *not* in fact run. However, we also know that if N is not prime, then regardless of which way the photon goes through the device we will observe the answer $|off\rangle|r = 0\rangle$ 100% of the time. Thus to find the photon emerge from the vertical output port of BS2 we must infer that N *is* prime. Thus, putting our two inferences together we are led to conclude that although the computer did *not* run, the number tested *is* prime! This is the counterfactual result. We have got something for free. The mere fact that if the computer had run it would have given the correct answer has allowed us to infer the correct answer without running the computer.

An alternative way of explaining how counterfactual results arise in quantum mechanics can be based on a different "interpretation" of quantum theory called the "Many Worlds Interpretation." In the Many Worlds view of quantum mechanics, when a quantum system must make a choice between two alternatives (such as whether a photon should traverse the upper or lower path after BS1) physical reality splits into two worlds; one in which the decision came out one way, and the other in which it came out the other way. For example, at BS1 in one world the photon went via the upper path and did not trigger the computer to turn on and in another world the photon went via the lower path and did turn the computer on. So in one world the computation was not done and r remained as $|r = 0\rangle$, whereas in the other world the computation was performed and r became set to $|r = 1\rangle$. To get the $|on\rangle|r = 0\rangle$ result we must be in the world in which the computation was *not* performed. However, at BS2, these two worlds partly remerged and we obtained some information about the computation that was performed in the other world.

It is largely a matter of philosophical taste as to which interpretation of quantum theory you prefer as there is no hard experimental evidence that allows us to distinguish between competing interpretations, although this does not prevent several physicists from having passionate opinions about their preferred interpretation.

Note that overall there is a 3/4 probability that the quantum computing scheme will correctly report a prime number. In two cases the computer actually had to run, but in the counterfactual case we obtained our knowledge that the number is prime without the computer running! The experimental setup has effectively given us extra computational power that is not available classically—the ability to compute in parallel worlds.

To sum up, by seeing $|on\rangle$ we *infer* the result of the computation to be $|r = 1\rangle$. But by reading the register and finding $|r = 0\rangle$, we determine that the computer was not in fact run (for, if it had, the output register must show $|r = 1\rangle$). Thus, with probability 1/4 we learn the correct result, and know it is correct without any computation having taken place!

It is too soon to tell whether counterfactual quantum computing will offer any advantage over regular quantum computing. Certainly, it does not solve the problem of decoherence and errors in quantum computers because the nonrunning quantum computer will inherit the errors it would have incurred if it had been run. Perhaps a better way to think of this is to believe that the computation was performed in some other parallel world in which all the usual error processes were at work. In fact, although the quantum computer is not actually run, it still needs to be capable of solving the desired problem if it were run. This will necessitate building fault-tolerance and error correction into the computer. As in the case of interaction-free measurement we can boost the probability of obtaining a counterfactual result arbitrarily close to 1.

Thus, in this chapter we have seen that not only does quantum computation offer "truth-without-proof" it also offers "truth-without-computation"! For many years we have grown used to our PCs telling us when it is safe to turn them off, after we are done computing. With quantum computers, it now seems safe to turn them off—and still compute!

Epilogue:
Quantum Technologies
in the Twenty-First Century

We are now at the end of our brief tour of quantum computing, and it is time to take stock of what we have learned and what we might expect in the first few decades of the twenty-first century.

The first tantalizing evidence of quantum phenomena came only 100 years ago. It took another quarter century before the first comprehensive quantum theory was developed, and the field has been growing in fascinating directions ever since.

The first results on quantum computing appeared less than two decades ago. Most notable were Paul Benioff's quantum description of a classical Turing machine, Richard Feynman's speculations on the relative efficiencies of classical versus quantum computers for simulating physical processes, and David Deutsch's models of true quantum Turing machines and quantum circuits. Thereafter, little else happened for the better part of a decade except for a handful of results demonstrating various arcane but important complexity separations between quantum computing and classical computing. The critical event that ignited interest in quantum computing was Peter Shor's discovery (building on work by Dan Simon) of an efficient quantum algorithm for factoring large composite integers. This was the first problem of practical significance for which a quantum algorithm was shown to exponentially better than any known classical algorithm. This led to a rapid influx of research funds for quantum computing from various government agencies, especially the National Security Agency (NSA), the Defense Advanced Research Projects Agency (DARPA), and the Army Research Office (ARO). It is a curious coincidence that the early work in quantum computing was inspired by military and security interests as was the early work in classical computing.

In addition to progress in quantum computing, an even more impressive development in quantum communications has occurred. In fact, by 1989 an experimental prototype of a quantum cryptography device was up and running. Since that time, quantum communications, especially quantum cryptography, and lately quantum teleportation, have matured into practical technologies.

For quantum computing and quantum communications to have reached their current levels of maturity in just over a decade of research effort has been remarkable. But the field has had, and still has, its critics. For the most part the critics argue that universal quantum computation will be impossible over arbitrarily many computational steps. Such criticism would have seemed far more damning five years ago than it does today. In the interim, scientists have discovered how to correct errors in a quantum computer without ever having to reveal the error explicitly. And the overhead in performing the error correction is not unreasonable. Moreover, we now know that if the reliability of the individual quantum gates used in a quantum circuit could be made to exceed 99.9999%, then it is possible to devise a hierarchical error correction scheme that can correct errors in quantum computations indefinitely, at least in principle. This even allows for the possibility of the error correction circuitry introducing new errors. Of course, in practice a 99.9999% reliability is still technologically demanding, but it is not completely out of the question.

Moreover, we would argue that it is too harsh to define the measure of success of quantum computing in terms of whether we can achieve arbitrary universal quantum computation. Quantum computers are simply not suited to all computational problems. So universal computation is not the true goal. Rather, the unique properties of quantum computers make them ideal tools for certain specialized applications, such as code-breaking and simulating quantum systems. In these applications there might be particular tricks for compiling quantum circuits down to manageable complexity that will allow progress to be made. Even rudimentary quantum computers that can control around 40–50 qubits would begin to allow us to solve certain problems, such a calculating eigenvalues of huge matrices, which are quite beyond the reach of conventional classical supercomputers.

Perhaps more surprising, is the realization that a quantum information-theoretic perspective on physical processes can shed new light on the foundations of physics itself. For example, quantum information theory might provide a deeper understanding of the origins of physical laws. It might also improve understanding of the inner workings of black holes. Possible new applications are appearing every month.

Beyond purely theoretical considerations, quantum computing has great potential technologically. Various kinds of quantum sensors appear to be feasible that require relatively modest advances in state-of-the-art quantum technology. For example, interaction-free measurement has already been adapted to perform rudimentary interaction-free imaging (White et al., 1998). In principle, this allows you to take a photograph of an object without any photons touching the object! This could have huge implications for medical imaging (Vaidman, 1996). Secondly, quantum-optical gyroscopes appear to be technologically within reach that should be a million times to one hundred million times more sensitive to rotations than conventional gyroscopes (Dowling, 1998). Such ultraprecise gyroscopes would allow point-to-point inertial navigation on the Earth to an unprecedented degree of accuracy. Finally, quantum cryptography has already advanced to the point where it can be commercialized. There are already proposals afoot to wire the financial district of London and the government agencies around Washington, D.C., with a quantum cryptographic network for guaranteed unbreakable communications (Ross, 1999).

In fact, if you take stock of the current trends it is as if we are in the midst of a second quantum revolution. The first quantum revolution was the discovery of the basic quantum effects such as wave–particle duality, Schrödinger's equation, quantization, and the Heisenberg Uncertainty Principle. Now, however, the focus is shifting to issues of multiparticle quantum mechanics, entanglement, and nonlocality. These newer, more "quantum" quantum effects have commercial applications that could affect all our lives.

Quantum computers require a drastic re-evaluation of the foundations of computer science and information theory. It is surprising to us that relatively few computer scientists seem to appreciate this. The prevailing attitude treats quantum computing as a passing fad. However, we believe that this is a shortsighted attitude. Quantum computers can perform certain computational tasks in exponentially fewer steps than any classical computer. Moreover, they facilitate unprecedented tasks such as teleporting quantum information and communicating with messages that betray the presence of eavesdropping. Technological progress has been modest so far, but when we understand how to implement quantum computing in solid-state quantum electronics, it is likely to develop considerably faster. Indeed, several patents have already been filed on core quantum computing and quantum communications technologies. If the next 75 years of quantum computing result in as many insights as did the first 75 years of quantum physics, then the future looks bright indeed.

References

Chapter 1

Feynman, R. P. "There's Plenty of Room at the Bottom," *Engineering and Science,* Vol. 23, pp. 22–36 (1960).

Hutcheson, G. D., and J. D. Hutcheson. "Technology and Economics in the Semiconductor Industry," *Scientific American,* January, pp. 54–62 (1996).

Keyes, R. W. "Miniaturization of Electronics and Its Limits," *IBM Journal of Research and Development,* Vol. 32, January, pp. 24–28 (1988).

Likharev, K. "Rapid Single-Flux Quantum Logic," *PhysComp '96,* New England Complex Systems Institute, pp. 194–195 (1996).

Malone, M. *The Microprocessor: A Biography,* TELOS/Springer-Verlag, New York, CA (1995).

Marand, C., and P. Townsend. "Quantum Key Distribution Over Distances as Long as 30 km," *Optics Letters,* Vol. 20, No. 16, 15 August, pp. 1695–1697 (1995).

Chapter 2

Barenco, A. "A Universal Two-Bit Gate for Quantum Computation," *Proceedings Royal Society of London, Series A,* Vol. 449, pp. 679–683 (1995).

Cleve, R., Ekert, A., Macchiavello, C., and M. Mosca. "Quantum Algorithms Revisited," *Proceedings Royal Society of London, Series A,* Vol. 454, pp. 339–354 (1997); also available at Los Alamos preprint archive, http://xxx.lanl.gov/archive/quant-ph/9708016.

DiVincenzo, D. "Two-Bit Gates Are Universal for Quantum Computation," *Physical Review A,* Vol. 51, pp. 1015–1022 (1995).

Ekert, A., and R. Jozsa. "Quantum Algorithms: Entanglement Enhanced Information Processing," *Philosophical Transactions of the Royal Society of London, Series A,* Vol. 356, 1769–1782 (1998); Proceedings of Royal Society Discussion Meeting "Quantum Computation: Theory and Experiment," held in November 1997 (1998); also available at Los Alamos preprint archive, http://xxx.lanl.gov/archive/quant-ph/9803072.

Chapter 3

Abrams, D., and S. Lloyd. "A Quantum Algorithm Providing Exponential Speed Increase for Finding Eigenvalues and Eigenvectors," http://xxx.lanl.gov/archive/quant-ph/9807070 (1998).

Appel, K., and W. Haken. "The Solution of the Four-Color-Map Problem," *Scientific American,* Vol. 237, No. 4, pp. 108–121 (1977).

Beals, R., Buhrman, H., Cleve, R., Mosca, M., and R. de Wolf. "Quantum Lower Bounds by Polynomials," *Proceedings of the 39th Annual Symposium on Foundations of Computer Science,* FOCS '98, Palo Alto, CA, November 8–11, pp. 352–361 (1998).

Benioff, P. "The Computer as a Physical System: A Microscopic Quantum-Mechanical Hamiltonian Model of Computers as Represented by Turing Machines," *Journal of Statistical Physics,* Vol. 22, pp. 563–591 (1980).

Bennett, C. "Logical Reversibility of Computation," *IBM Journal of Research and Development,* Vol. 17, pp. 525–532 (1973).

Bernstein, E., and U. Vazirani. "Quantum Complexity Theory," *Proceedings of the 25th Annual ACM Symposium on the Theory of Computing,* pp. 11–20 (1993).

Berthiaume, A., and G. Brassard. "The Quantum Challenge to Complexity Theory," *Proceedings of the 7th IEEE Conference on Structure in Complexity Theory,* pp. 132–137 (1992).

Berthiaume, A., and G. Brassard. "Oracle Quantum Computing," *Journal of Modern Optics,* Vol. 41, No. 12, December, pp. 2521–2535 (1994).

Brassard, G., Hoyer, P., and A. Tapp. "Quantum Counting," *Proceedings of Automata, Languages, and Programming, 25th International Colloquium, ICALP '98,* Aalborg, Denmark, 13–17 July, pp. 820–831 (1998).

Cerf, N., Grover, L., and C. Williams. "Nested Quantum Search and NP-Complete Problems," Los Alamos preprint archive, http://xxx.lanl.gov/archive/quant-ph/9806078 (1998).

Crandall, R. *Topics in Advanced Scientific Computation,* TELOS/Springer Verlag, New York, pp. 125–126 (1996).

Deutsch, D. "Quantum Theory, the Church–Turing Principle, and the Universal Quantum Computer," *Proceedings Royal Society London, Series A,* Vol. 400, pp. 97–117 (1985).

Deutsch, D., and R. Jozsa. "Rapid Solution of Problems by Quantum Computation," *Proceedings Royal Society London, Series A,* Vol. 439, pp. 553–558 (1992).

Durr, C., and P. Hoyer. "A Quantum Algorithm for Finding the Minimum," Los Alamos preprint archive, http://xxx.lanl.gov/archive/quant-ph/9607014 (1996).

Feynman, R. "Simulating Physics with Computers," *International Journal of Theoretical Physics,* Vol. 21, Nos. 6/7, pp. 467–488 (1982).

Fijany, A., and C. Williams. "Quantum Wavelet Transforms: Fast Algorithms and Complete Circuits," Spinger-Verlag Lecture Notes in Computer Science, Volume

1509, Springer-Verlag, Heidelberg (1999); also available as http://xxx.lanl.gov /abs/quant-ph/9809004.

Gill, J. "Computational Complexity of Probabilistic Turing Machines," *SIAM Journal of Computing,* Vol. 6, No. 4, December, pp. 675–695 (1977).

Gorenstein, D. *Finite Simple Groups,* Plenum, New York (1982).

Grover, L. "A Fast Quantum-Mechanical Algorithm for Database Search," *Proceedings of the 28th Annual ACM Symposium on the Theory of Computing,* pp. 212–219 (1996).

Grover, L. "A Fast Quantum-Mechanical Algorithm for Estimating the Median," AT&T Bell Labs preprint (1996).

Grover, L. "Quantum Mechanics Helps in Searching for a Needle in a Haystack," *Physical Review Letters,* Vol. 79, pp. 325–328 (1997).

Grover, L. "Quantum Computers Can Search Rapidly by Using Almost Any Transformation," *Physical Review Letters,* Vol. 80, pp. 4329–4333 (1998).

Hopcroft, J. "Turing Machines," *Scientific American,* May, pp. 86–98 (1984).

Jozsa, R. "Characterizing Classes of Functions Computable by Quantum Parallelism," *Proceedings Royal Society London, Series A,* Vol. 435, pp. 563–574 (1991).

Kitaev, A. "Quantum Measurements and the Abelian Stabiliser Problem," Los Alamos preprint archive, http://xxx.lanl.gov/archive/quant-ph/9511026 (1995).

Lenstra, A., Manasse, M., and J. Pollard. "The Number Field Sieve," *Proceedings of the 22nd ACM Symposium on the Theory of Computing,* pp. 564–572 (1990).

Lloyd, S. "Quantum-Mechanical Computers and Uncomputability," *Physical Review Letters,* Vol. 71, pp. 943–946 (1993).

Lloyd, S. "Universal Quantum Simulators," *Science,* Vol. 273, 23 August, pp. 1073–1078 (1996).

Peres, A., and W. Zurek. "Is Quantum Theory Universally Valid?," *American Journal of Physics,* Vol. 50, September, pp. 807–810 (1982).

Peres, A. "Einstein, Gödel, Bohr," *Foundations of Physics,* Vol. 15, pp. 201–205 (1985).

Shapiro, S., ed. "Church's Thesis," *Encyclopedia of Artificial Intelligence,* John Wiley & Sons, New York, pp. 99–100 (1990).

Shor, P. "Algorithms for Quantum Computation: Discrete Logarithms and Factoring," *Proceedings 35th Annual Symposium on Foundations of Computer Science,* pp. 124–134 (1994).

Silverman, R. "The Multiple Polynomial Quadratic Sieve," *Mathematics of Computing,* Vol. 48, pp. 329–338 (1987).

Simon, D. "On the Power of Quantum Computation," *Proceedings of the 35th Annual IEEE Symposium on Foundations of Computer Science,* pp. 116–123 (1994).

Turing, A. "On Computable Numbers with an Application to the Entscheidungsproblem," *Proceedings of the London Mathematical Society,* Vol. 42, pp. 230–265 (1937); erratum in Vol. 43, pp. 544–546 (1937).

van Dam, W. "Quantum Oracle Interrogation: Getting All the Information for Almost Half the Price", *Proceedings of the 39th Annual Symposium on Foundations of Computer Science,* FOCS '98, Palo Alto, CA, November 8–11, pp. 362–367 (1998).

Wiles, A. "Modular Elliptic Curves and Fermat's Last Theorem," *Annals of Mathematics,* Vol. 141. pp. 443–551 (1995).

Williams, C., and T. Hogg. "Exploiting the Deep Structure of Constraint Problems," *Artificial Intelligence Journal,* Vol. 70, pp. 73–117 (1994).

Winograd, T. *Understanding Natural Language,* Academic Press, New York (1972).

Yao, A. "Quantum Circuit Complexity," *Proceedings of the 34th IEEE Symposium on Foundations of Computer Science,* IEEE Computer Society Press, Los Alamitos, CA, pp. 352–360 (1993).

Chapter 4

Brassard, G., Hoyer, P., and A. Tapp. "Quantum Counting," *Proceedings of Automata, Languages, and Programming, 25th International Colloquium, ICALP '98,* Aalborg, Denmark, 13–17 July, pp. 820–831 (1998).

Lenstra, A., and H. Lenstra. "The Development of the Number Field Sieve," *Lecture Notes in Mathematics,* Vol. 1554, Springer-Verlag, New York (1993).

Oakley, B. British Computer Society, personal communication.

Rivest, R., Shamir, A., and L. Adelman. "A Method for Obtaining Digital Signatures and Public Key Cryptosystems," *Communications of the ACM,* Vol. 21, pp. 120–126 (1978).

Shor, P. "Algorithms for Quantum Computation: Discrete Logarithms and Factoring," *Proceedings 35th Annual Symposium on Foundations of Computer Science,* pp. 124–134 (1994).

Simon, D. "On the Power of Quantum Computation," *Proceedings of the 35th Annual IEEE Symposium on Foundations of Computer Science,* pp. 116–123 (1994).

Vazirani, U. Quotation from a newspaper article by Tom Siegfried, Science Editor of the *Dallas Morning News,* after a 1994 conference sponsored by the Santa Fe Institute, Los Alamos National Laboratory, and the University of New Mexico.

Welsh, D. *Codes and Cryptography,* Oxford University Press, Oxford, p. 183 (1988).

Chapter 5

Abbot, P., ed. "Tricks of the Trade: Random Numbers," *Mathematica Journal,* Vol. 5, Issue 1, pp. 20–21 (1995).

Abrams, D., and C. Williams. "Fast Computation of Integrals Using a Quantum Computer," submitted to *Physical Review A* (1999).

Beardsley, T. "Rebottling the Nuclear Genie," *Scientific American,* May, pp. 34–35 (1998).

Berdnikov, A., Turtia, S., and A. Companger. "A MathLink Program for High-Quality Random Numbers," *Mathematica Journal,* Vol. 6, Issue 3, pp. 65–69 (1996).

Chaitin, G. "Algorithmic Information Theory," *IBM Journal of Research and Development,* Vol. 21, July, pp. 350–359 (1977).

Chaitin, G. "Gödel's Theorem and Information," *International Journal of Theoretical Physics,* Vol. 22, pp. 941–954 (1982); also as http://www.umcs.edu/~chaitin/georgia.html.

Deutsch, D. "Quantum Theory, the Church–Turing Principle, and the Universal Quantum Computer," *Proceedings Royal Society London, Series A,* Vol. 400, pp. 97–117 (1985).

Ferrenberg, A., Landau, D., and Y. Wong. "Monte Carlo Simulations: Hidden Errors from 'Good' Random Number Generators," *Physical Review Letters,* Vol. 69, pp. 3382–3384 (1992).

Feynman, R. "Simulating Physics with Computers," *International Journal of Theoretical Physics,* Vol. 21, Nos. 6/7, pp. 467–488 (1982).

Gardner, M. "Mathematical Games: The Random Number Omega Bids Fair to Hold the Mysteries of the Universe," *Scientific American,* November, pp. 20–30 (1979).

Gibbs, W. "Taking Aim At Tumors," *Scientific American,* May, pp. 17–20 (1998).

Gleick, J. *Chaos,* Cardinal, London (1987).

Hille, B. *Ionic Channels of Excitable Membranes,* 2nd ed., Sinauer Associates, Sunderland, MA, p. 389 (1992).

Hughes, R., Alde, D., Dyer, P., Luther, G., Morgan, G., and M. Schauer. "Quantum Cryptography," *Contemporary Physics,* Vol. 36, pp. 149–163 (1995).

James, F. "A Review of Pseudorandom Number Generators," *Computer Physics Communications,* Vol. 60, pp. 329–344 (1990).

Marsaglia, G. "Random Numbers Fall Mainly in the Planes," *Proceedings of the National Academy of Sciences,* Vol. 61, September, pp. 25–28 (1968).

Metropolis, N., Rosenbluth, A., Rosenbluth, M., Teller, A., and E. Teller. "Equation of State Calculations by Fast Computing Machines," *Journal of Chemical Physics,* Vol. 21, pp. 1087–1092 (1953).

Park, S., and K. Miller. "Random Number Generators: Good Ones Are Hard to Find," *Communications of the ACM,* Vol. 31, pp. 1192–1201 (1988).

Pour-El, M., and I. Richards. "Noncomputability in Models of Physical Phenomena" *International Journal of Theoretical Physics,* Vol. 21, pp. 553–555 (1982).

Schroeder, M. *Fractals, Chaos, Power Laws,* W. H. Freeman, New York (1990).

Shannon, C. "Computers and Automata," *Proceedings of the I. R. E.,* Vol. 41, October, pp. 1235–1241 (1953).

Svozil, K. "Quantum Algorithmic Information Theory," Los Alamos preprint archive http://xxx.lanl.gov/quant-ph/951005 (1999).

Traub, J., and H. Wozniakowski. "Breaking Intractability," *Scientific American,* January, pp. 102–107 (1994).

Wiesner, S. "Simulations of Many-Body Quantum Systems by a Quantum Computer," Los Alamos preprint archive, http://xxx.lanl.gov/archive/quant-ph/9603028 (1996).

Zak, M., and C. Williams. "Quantum Recurrent Networks for Simulating Stochastic Processes," *Springer-Verlag Lecture Notes in Computer Science,* Volume 1509, Springer-Verlag, Heidelberg (1999).

Chapter 6

Bennett, C., and G. Brassard. "The Dawn of a New Era for Quantum Cryptography: The Experimental Prototype is Working!" *SIGACT News,* Vol. 20, Fall, pp. 78–82 (1989).

Bennett, C., Bessette, F., and G. Brassard. "Experimental Quantum Cryptography," *Lecture Notes in Computer Science,* Vol. 473, Springer-Verlag, Berlin, pp. 253–265 (1991).

Bennett, C., Bessette, F., Brassard, G., Salvail, L., and J. Smolin. "Experimental Quantum Cryptography," *Journal of Cryptology,* Vol. 5, pp. 3–28 (1992).

Bennett, C., Brassard, G. and A. Ekert. "Quantum Cryptography," *Scientific American,* October, pp. 50–57 (1992).

Buttler, W., Hughes, R., Kwiat, P., Luther, G., Morgan, G., Nordholt, J., Peterson, C., and C. Simmons. "Free-Space Quantum Key Distribution," Los Alamos preprint archive, http://xxx.lanl.gov/archive/quant-ph/9801006 (1998).

Buttler, W., Hughes, R., Kwiat, P., Lamoreaux, S., Luther, G., Morgan, G., Nordholt, J., Peterson, C., and C. Simmons. "Practical Free-Space Quantum Key Distribution Over 1 Kilometer," Los Alamos preprint archive, http://xxx.lanl.gov/archive/quant-ph/9805071 (1998).

Deutsch, D. "Quantum Communication Thwarts Eavesdroppers," *New Scientist,* December 9, pp. 25–26 (1989).

Ekert, A. "Quantum Cryptography Based on Bell's Theorem," *Physical Review Letters,* Vol. 67, pp. 661–663 (1991).

Ekert, A., Rarity, J., Tapster, P., and G. Palma. "Practical Quantum Cryptography Based on Two-Photon Interferometry," *Physical Review Letters,* Vol. 69, 31 August, pp. 1293–1295 (1992).

Franson, J., and H. Ilves. "Quantum Cryptography Using Optical Fibers," *Applied Optics,* Vol. 33, pp. 2949–2954 (1995).

Hecht, E., and A. Zajac. *Optics,* Addison-Wesley, Reading, MA, p. 228 (1974).

Hughes, R., Alde, D., Dyer, P., Luther, G., Morgan, G., and M. Schauer. "Quantum Cryptography," *Contemporary Physics,* Vol. 36, pp. 149–163 (1995).

Hughes, R., James, D., Knill, E., Laflamme, R., and A. Petschek. "Decoherence Bounds on Quantum Computation with Trapped Ions," *Physical Review Letters,* Vol. 77, pp. 3240–3243 (1996).

Hughes, R., Buttler, W., Kwiat, P., Luther, G., Morgan, G., Nordholt, J., Peterson, C., and C. Simmons. "Secure Communications Using Quantum Cryptography," *Proceedings of SPIE,* Vol. 3076, pp. 2–11 (1997).

Marand, C., and P. Townsend "Quantum Key Distribution Over Distances as Long as 30 km," *Optics Letters,* 15 August, Vol. 20, No. 16, pp. 1695–1697 (1995).

Muller, A., Zbinden, H., and N. Gisin. "Quantum Cryptography Over 23 km in Installed Under-Lake Telecom Fibre," *Europhysics Letters,* Vol. 33, pp. 335–339 (1996).

Townsend, P., Rarity, J., and P. Tapster. "Single Photon Interference in 10 km-Long Optical Fibre Interferometer," *Electronics Letters,* Vol. 29, 1 April, pp. 634–635 (1993).

Townsend, P., Rarity, J., and P. Tapster. "Enhanced Single Photon Fringe Visibility in a 10 km-Long Prototype Quantum Cryptography Channel," *Electronics Letters,* Vol. 29, July, pp. 1291–1293 (1993a).

Chapter 7

Aspect, A., Dalibard, J., and G. Roger. "Experimental Test of Bell's Inequalities Using Time-Varying Analyzers," *Physical Review Letters,* Vol. 49, pp. 1804–1807 (1982).

Bennett, C., Brassard, G., Crepeau, C., Jozsa, R., Peres, A., and W. Wootters. "Teleporting an Unknown Quantum State via Dual Classical and Einstein–Podolsky–Rosen Channels," *Physical Review Letters,* Vol. 70, pp. 1895–1899 (1993).

Boschi, D., Branca, S., De Martini, F., Hardy, L., and S. Popescu. "Experimental Realization of Teleporting an Unknown Pure Quantum State via Dual Classical and Einstein–Podolski–Rosen Channels," *Physical Review Letters,* Vol. 80, pp. 1121–1125 (1998).

Clauser, J., and A. Shimony. "Bell's Theorem: Experimental Tests and Implications," *Reports on Progress in Physics,* Vol. 41, pp. 1881–1927 (1978).

Chapter 8

Barenco, A., Berthiaume, A., Deutsch, D., Ekert, A., Jozsa, R., and C. Macchiavello. "Stabilisation of Quantum Computations by Symmetrization," Los Alamos preprint archive, http://xxx.lanl.gov/archive/quant-ph/9604028 (1996).

DiVincenzo, D. "Quantum Computation," *Science,* Vol. 270, 13 October, pp. 255–261 (1995).

Ekert, A., and C. Macchiavello. "Quantum Error Correction for Communication," *Physical Review Letters,* Vol. 77, pp. 2585–2588 (1996).

Joos, E., and H. Zeh. "The Emergence of Classical Properties Through Interaction with the Environment," *Zeitschrift fur Physik B,* Vol. 59, pp. 223–243 (1985).

Kitaev, A. "Fault-Tolerant Quantum Computation by Anyons," Los Alamos preprint archive http://xxx.lanl.gov/archive/quant-ph/9707021 (1997).

Laflamme, R., Miquel, C., Paz, P., and W. Zurek. "Perfect Quantum Error-Correcting Code," *Physical Review Letters,* Vol. 77, pp. 198–201 (1996).

Landauer, R. "Is Quantum Mechanics Useful?" *Philosophical Transactions Royal Society,* Series A, Vol. 353, pp. 367–376 (1995).

Preskill, J. "Fault-Tolerant Quantum Computation," Los Alamos preprint archive http://xxx.lanl.gov/archive/quant-ph/9712048 (1997a).

Preskill, J. "Reliable Quantum Computers," Los Alamos preprint archive http://xxx.lanl.gov/archive/quant-ph/9705031 (1997b).

Shor, P. "Scheme for Reducing Decoherence in Quantum Computer Memory," *Physical Review A,* Vol 52, October, pp. R2493–R2496 (1995).

Unruh, W. "Maintaining Coherence in Quantum Computers," *Physical Review A,* Vol. 51, pp. 992–997 (1995).

Chapter 9

Braunstein, S., Caves, C., Jozsa, R., Linden, N., Popescu, S., and R. Schack. "Separability of Very Noisy Mixed States and Implications for NMR Quantum Computing," Los Alamos preprint archive http://xxx.lanl.gov/quant-ph/9811018 (1998).

Chuang, I., Gershenfeld, N., and M. Kubinec. "Experimental Implementation of Fast Quantum Searching," *Physical Review Letters,* Vol. 80, pp. 3408–3412 (1998). See also: Seife, C. "Quantum Leap," *New Scientist,* April 18, p. 10 (1998); http://feynman.stanford.edu/qcomp, and http://www.sciam.com/1998 /0698issue/0698gershenfeld.html.

Chuang, I., Vandersypen, L., Zhou, X., Leung, D., and S. Lloyd. "Experimental Realization of a Quantum Algorithm," Los Alamos preprint archive http:// xxx.lanl.gov/quant-ph/9801037 (1998a).

Cirac, J., and P. Zoller. "Quantum Computations with Cold Trapped Ions," *Physical Review Letters,* Vol. 74, pp. 4091–4094 (1995).

Cory, D., Fahmy, A., and T. Havel. "Nuclear Magnetic Resonance Spectroscopy: An Experimentally Accessible Paradigm," Los Alamos preprint archive http://xxx.lanl.gov/quant-ph/9709001 (1997).

DiVincenzo, D. "Real and Realistic Quantum Computers," *Nature,* Vol. 393, 14 May, pp. 113–114 (1998).

Gershenfeld, N., and I. Chuang. "Quantum Computing with Molecules," *Scientific American,* June, pp. 66–71 (1998).

Hughes, R., James, D., Knill, E., Laflamme, R., and A. Petschek. "Decoherence Bounds on Quantum Computation with Trapped Ions," Los Alamos preprint archive http://xxx.lanl.gov/quant-ph/9604026 (1996).

Jewell, J., Harbison, J., and A. Scherer. "Microlasers," *Scientific American,* November, pp. 86–94 (1991).

Jones, J., and M. Mosca. "Implementation of a Quantum Algorithm to Solve Deutsch's Problem on a Nuclear Magnetic Resonance Computer," Los Alamos preprint archive http://xxx.lanl.gov/quant-ph/9801027 (1998).

Kane, B. "A Silicon-Based Nuclear Spin Quantum Computer," *Nature,* Vol. 393, 14 May, pp. 133–137 (1998).

Lloyd, S. "A Potentially Realizable Quantum Computer," *Science,* Vol. 261, 17 September, pp. 1569–1571 (1993).

Monroe, C., Meekhof, D., King, B., Itano, W., and D. Wineland. "Demonstration of a Fundamental Quantum Logic Gate," *Physical Review Letters,* Vol. 75, No. 25, pp. 4714–4717 (1995).

Preskill, J. "Quantum Computing: Pro and Con," Los Alamos preprint archive http://xxx.lanl.gov/quant-ph/9705032 (1997).

Privman, V., Vagner, I., and G. Kventsel. "Quantum Computation in Quantum-Hall Systems," Los Alamos preprint archive http://xxx.lanl.gov/quant-ph /9707017 (1997).

Steane, A. "The Ion Trap Quantum Information Processor," Los Alamos preprint archive http://xxx.lanl.gov/archive/quant-ph/9608011 (1996).

Teich, W., Obermayer, K., and G. Mahler. "Structural Basis for Multistationary Quantum Systems II: Effective Few-Particle Dynamics," *Physical Review B,* Vol. 37, No. 14, pp. 8111–8120 (1988).

Turchette, Q., Hood, C., Lange, W., Mabuchi, H., and H. Kimble. "Measurement of Conditional Phase Shifts for Quantum Logic," *Physical Review Letters,* Vol. 75, No. 25, pp. 4710–4713 (1995).

Chapter 10

Cerf, N., Adami. C., and P. Kwiat. "Optical Simulation of Quantum Logic," http://xxx.lanl.gov/quant-ph/9706022 (1997).

Dicke, R. "Interaction-Free Quantum Measurements: A Paradox?" *American Journal of Physics,* Vol. 49, November, pp. 925–930 (1981).

Elitzur, A., and L. Vaidman. "Quantum-Mechanical Interaction-Free Measurements," *Foundations of Physics,* Vol. 23, pp. 987–997 (1993).

Jozsa, R. "Quantum Effects in Algorithms," in *Quantum Computing and Quantum Communications,* C. P. Williams, ed., Springer-Verlag, Lecture Notes in Computer Science, Volume 1509; also as http://xxx.lanl.gov/quant-ph/9805086 (1998).

Kwiat, P., Weinfurter, H., Herzog, T., and A. Zeilinger. "Interaction-Free Measurement," *Physical Review Letters,* Vol. 74, pp. 4763–4766 (1995).

Kwiat, P., White, A., Mitchell, J., Nairz, O., Weihs, G., Weinfurter, H., and A. Zeilinger. Personal communication (1999).

Epilog

Dowling, J. "Correlated Input-Port, Matter-Wave Interferometer: Quantum-Noise Limits to the Atom-Laser Gyroscope," *Physical Review A,* Vol. 57, pp. 4736–4746 (1998).

Ross, C. Quantum Electronic Devices PLC (a company commercializing quantum computing technology), private communication (1999).

Vaidman, L. "Interaction-Free Measurements," http://xxx.lanl.gov/quant-ph /9610033 (1996).

White, A., Mitchell, J., Nairz, O., and P. Kwiat. "Interaction-Free Imaging," http:// xxx.lanl.gov/quant-ph/9803060 (1998).

Index

ST. JOHN FISHER COLLEGE LIBRARY

0 1220 0067331 1

QA 76.889 .W55 2000
Williams, Colin P.
Ultimate zero and one